AEROTHERMODYNAMICS OF LOW PRESSURE STEAM TURBINES AND CONDENSERS

AEROTHERMODYNAMICS OF LOW PRESSURE STEAM TURBINES AND CONDENSERS

Edited by

M. J. Moore

Central Electricity Research Laboratory
Leatherhead, England

C. H. Sieverding

von Karman Institute for Fluid Dynamics
Rhode-Saint-Genèse, Belgium

 A von Karman Institute Book

HEMISPHERE PUBLISHING CORPORATION
A subsidiary of Harper & Row, Publishers, Inc.

Washington   New York   London

DISTRIBUTION OUTSIDE NORTH AMERICA

SPRINGER–VERLAG

Berlin   Heidelberg   Paris   New York   Tokyo   London

AEROTHERMODYNAMICS OF LOW PRESSURE STEAM TURBINES AND CONDENSERS

1 2 3 4 5 6 7 8 9 0   B R B R   8 9 8 7 6

Library of Congress Cataloging-in-Publication Data

Aerothermodynamics of low pressure steam turbines and
condensers.

"A von Karman Institute book."
Includes bibliographies and index.
1. Steam-turbines—Aerodynamics. 2. Steam-turbines—
Thermodynamics. 3. Condensers (Steam)—Aerodynamics.
4. Condensers (Steam)—Thermodynamics. I. Moore,
M. J. II. Sieverding, C. H., date.
TJ737.A35 1986   621.1'65   86-18357
ISBN 0-89116-446-4   Hemisphere Publishing Corporation

DISTRIBUTION OUTSIDE NORTH AMERICA:
ISBN 3-540-17086-3 Springer-Verlag   Berlin

D
621·165
AER

Contents

Preface

In the field of large scale power generation the steam turbine occupies a central position whether the energy source is a fossil fueled boiler, a gas cooled nuclear reactor, a light or heavy water reactor, a fast reactor, or, looking ahead, even a nuclear fusion reactor. The high cost of fossil fuels, however, and the high capital cost of nuclear systems makes it increasingly important to convert the liberated heat-energy to mechanical power as efficiently as possible. This volume describes key developments in the quest for higher turbine efficiency.

The 1970s saw the expansion of turbine size, reaching unit outputs of up to 1300 MW. In contrast, the downturn in world economic growth in the 1980s has resulted in a sharp reduction in orders for new machines and many utilities are taking steps to extend the operating life of older turbines. An essential element of the life extension exercise is the retrofitting of new components, using the latest design theories, to improve efficiency. Such schemes have been found highly cost-beneficial.

Prime candidates for retrofitting are the low-pressure turbine and condenser where, as unit sizes have increased, the aerothermodynamic design problems have been more difficult than in other parts of the machine. For these reasons the von Karman Institute has brought together experts in the field of low pressure turbine and condenser research to provide advanced Lecture Series in these subjects. This volume is a selection of edited lectures from these Series. The lecturers, from Europe and the USA, are specialists in their particular fields of research and development and this book is intended to provide students, researchers, and turbine plant designers with a view of the improvements in knowledge and techniques in recent years. Particularly significant, for example, are the emergence of theories for viscous compressible flow and the capability to measure steam wetness fraction, as described in Chapters 3 and 6, respectively, which must lead to further advances in turbine performance in the future.

By including the fluid mechanics of the turbine and condenser, this publication complements the previous von Karman Institute book "Two Phase Steam Flow in Turbines and Separators" edited by M. J. Moore and C. H. Sieverding (Hemisphere, 1976) and adds significantly to the treatment of wet steam flow in turbines.

We thank the following organizations for their cooperation: BBC Brown Boveri & Cie, Central Electricity Research Laboratories (CEGB), Kraftwerk Union AG, Westinghouse Electric Corporation, and Cambridge University. We also would like to express our gratitude to Mrs. N. V. Toubeau, Mrs. F. Postijns, Miss L. Klinkenbergh, Mr. J.–C. Lobet, and Mr. M. Blockx for their technical assistance in the preparation of this book.

M. J. Moore
C. H. Sieverding

The Authors

B. J. DAVIDSON is a Project Leader in the Thermodynamics Section at the CEGB research laboratories at Leatherhead, UK. He provides, in Section 8.1, a detailed description of heat transfer in large steam condensers and the modern computational techniques developed for tube-bundle design.

W. N. DAWES shows, in Section 3.2, the impressive results emerging from the new theoretical techniques for determining 2-D and 3-D viscous compressible flows through turbine blading. Dr. Dawes is making important contributions in this field during his work as a Lecturer in the Engineering Sciences Department of Cambridge University, UK.

J. D. DENTON is Director of the Whittle Laboratory of Cambridge University, UK. He describes inviscid flow methods and their application to the analysis of flows in transonic turbine blading, in Section 3.1. Dr. Denton is known internationally for his development of through-flow and time-marching computational techniques.

D. H. EVANS is a Fellow Engineer with Westinghouse Electric Corporation of Orlando, Florida, USA. He contributes Section 2.2 on matrix methods of computation of turbine flows and also Section 4.1 on experimental techniques and equipment for turbine blading development.

H. KELLER is Head of the Steam Turbine Thermodynamics Section at Kraftwerk Union, Mülheim (Ruhr), FRG. In Chapter 7 he introduces the particular complex flow fields occurring in turbine exhaust systems and discusses ways of minimizing the potentially high aerodynamic losses which can arise in these components.

H. V. LANG provides in Section 8.2 the manufacturers' approach to the development of large condensers and surveys the range of designs in the field. In his position as Chief Engineer at Brown Boveri Corporation, North Brunswick, USA, he describes the philosophy behind the well-known BBC 'church window' tube-bundle arrangement.

A. ACCORNERO and L. MARETTO make an extremely valuable contribution to the literature by providing details, in Section 5.2, of extensive flow field measurements in the final stages of an efficient 330 MW turbine. Mr. Accornero and Mr. Maretto are respectively in charge of the Turbomachinery Laboratory and of Turbine R&D at Ansaldo Componenti, S.p.A., Genoa, Italy.

M. J. MOORE is Programme Manager for turbine research at the CEGB research laboratories at Leatherhead, UK. His Section 5.1 describes recent advances in flow measurement techniques which are adding significantly to the knowledge of loss mechanisms in turbines.

A. ROEDER is Head of the Turbine and Generator Design Department at Brown Boveri and Cie, Baden, Switzerland. His Chapter 1 describes the mechanical and economic constraints encountered in the design of a large, highly efficient steam turbine.

C. H. SIEVERDING of the von Karman Institute, Rhode St. Genèse, Belgium, has 20 years of experience in testing transonic steam turbine cascades. In Section 4.2 he describes the complex aerodynamics of blade sections of the long, highly twisted rotor blades in the final stage of a modern L.P. turbine.

J. SNOECK carried out his research on wet steam flow through turbine blading during his stay as a post-graduate student at the von Karman Institute, before taking his present post as Scientific Advisor at the "Institut pour l'Encouragement de la Recherche Scientifique dans l'Industrie et l'Agriculture" (I.R.S.I.A.), Belgium. He describes the wet steam flow through the tip section of a final stage rotor blade in Section 6.2.

M. TROILO of the University of Genoa, Italy, has been developing for many years computer codes for the design and analysis of steam turbines. In Section 2.1 he describes the development of the streamline curvature through-flow theory which has played a central role in the development of modern steam turbines.

J. B. YOUNG is a Lecturer in the Engineering Sciences Department of Cambridge University, UK. His research field includes the study of wet-steam flow and in Section 6.1 he describes the latest theoretical developments for computing nucleation, droplet growth, and wet-steam losses in turbines.

AEROTHERMODYNAMICS OF LOW PRESSURE STEAM TURBINES AND CONDENSERS

CHAPTER 1

The Design of Modern Low Pressure Steam Turbines

A. Roeder

1.1 INTRODUCTION

Today's high cost of energy places great demands on the economics of electricity generation. Thus the successful design of a modern low pressure (LP) steam turbine is measured by criteria which govern the economics of this power plant component, in particular the criteria of availability, efficiency and capital cost. These criteria must be taken into account simultaneously in the LP steam turbine design. For example measures taken solely to improve the turbine efficiency may be inadvisable. Their consequences for turbine reliability must be considered and possible cost measures must be compared with potential efficiency gains using a realistic parity factor.

The simultaneous consideration of the above mentioned criteria, when designing the LP turbine, requires the close cooperation of the engineers responsible for the mechanical and the aerodynamic designs. To judge properly the design features of the product with respect to reliability, a constant feedback of operating experience from existing power plants is also required.

The efficiency of modern HP and IP steam turbines can be predicted very accurately. It is mainly a function of the flow losses in the blading. Those have been established in systematic experiments in the fluid mechanics laboratories for all relevant geometric dimensions and flow conditions. Different HP and IP turbine designs can therefore be reliably compared with respect to cost and efficiency.

The economic optimization of the LP turbine design is not always so successful. Wet steam and three-dimensional flow patterns make complete and accurate measurements difficult. The high costs of systematic tests with LP turbines have until now also prevented the accumulation of data which would permit an accurate efficiency analysis of the LP turbine alternatives. In addition, the LP efficiency is greatly affected by the losses in the exhaust casing. To date, these losses have hardly been amenable to theoretical treatment, so that an empirical evaluation based on a few measuring points was all that was possible. Thus, even today, analytical experimentally verified procedures are not available which would permit the economically optimum design of LP turbines. We continue to rely on the experience of engineers who are familiar with the complexity of the problem and can bridge the gaps with reasonable assumptions.

1.2 THE STRUCTURAL DESIGN OF THE LP TURBINE

Only by understanding the design features of the LP turbines is it possible to find reasonable concepts for the aerodynamic and mechanical designs. In this chapter the design of modern LP turbines is illustrated by two examples.

FIGURE 1.2.1 LP turbine 3600 min^{-1}

FIGURE 1.2.2 Shop assembly of the LP rotor (last stage blades not yet fitted)

1.2.1 3600 RPM LP Turbine

Figure 1.2.1 shows a 3600 RPM LP turbine of the type being operated in numerous North American power plants with outputs up to 1300 MW. The main components of this double-flow turbine consist of the outer casing, the inner casing with the stationary blade carriers, the blading, and the rotor. Each flow has an exhaust area of 6 m^2. The steam passes through thermally elastic steam penetrations, which are connected via expansion joints to the outer casing, into the inlet portion of the inner casing, and from there to the two opposed blading flows. After the blading, the steam flows through an annular, short diffusor into the exhaust space of the outer casing, and from there downwards to the condenser.

Both inner and outer casings are welded structures. Cast stationary blade carriers are inserted into the inner casing. The casings and blade carriers are horizontally split at the height of the turbine axis and are bolted at their flanges (Fig. 1.2.2). The outer casing rests on the foundation over the entire length of its lower half, on laterally attached supports. The longitudinal positioning with respect to the bearing pedestal or the generator is achieved by means of keys. The inner casing is supported at the joint face within the outer casing and axially fixed relative to the outer casing at the inlet portion. Sliding keys permit the free expansion of the casing in all directions. The bolted stationary blade carriers are inserted in the inner casing where they are axially fixed. Their free longitudinal and lateral expansion is assured.

Each turbine flow has 5 stages. The first 3 stages are of the reaction type with symmetrical constant section, shrouded blade profiles. Clearance losses are completely avoided in the first stationary row by mounting the blades in the inlet guide vane. The two last moving blade rows have free-standing, tapered blades which are twisted to match the three-dimensional flow. The stationary blades are made from plate steel and are cast-in and slightly profiled.

The LP rotor (Fig. 1.2.1) is welded together from forged discs and shaft-ends. At each end of the body of the rotor there are holes for the balancing weights. Rebalancing without opening the casing is possible. Axial gland seals, at the rotor penetrations, prevent the escape of steam. The seals are designed as labyrinths (Fig. 1.2.3) with radially arranged elastic ring segments, to avoid damage from rubbing. The axial thrust is equalized by the double-flow blading, in spite of the unsymmetrical extraction of steam for the feed heaters to such an extent that no dummy piston is required.

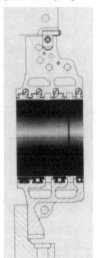

To prevent implosion as a result of the atmospheric pressure on the LP casing, internal supports are necessary. This should, however, be kept to a minimum to avoid flow losses. Changes in the vacuum, caused, e.g. by seasonal variations in the cooling water temperature, result in stress changes and lead to additional deformations of the endwalls.

This, in turn may cause problems with the gland seals if they are mounted directly upon the endwalls. The small radial clearance in the glands may be insufficient to accept the deformation of the shell. Thus in large LP turbines it is preferable to attach the gland seals to the bearing pedestals which are separated from the outer casing.

FIGURE 1.2.3 LP gland seals

The LP inner casing is subjected to relatively large pressure and thermal stresses. Excessive stresses can be avoided only by a very flexible design. The self-supporting blade carriers are installed in a thermally flexible way. Modern shell-theory calculation methods can solve the complex stress and

and deformation problems, so that dangerous rubbing of the rotor or internal leakages can be avoided.

1.2.2 1800 RPM LP Turbine

Figure 1.2.4 shows a modern half speed LP turbine, used in North American nuclear power plants of the 900 to 1300 MW class. The exhaust area of the blading is 16.4 m^2 per flow, i.e., 2.7 times the area of the previously shown full-speed turbine.

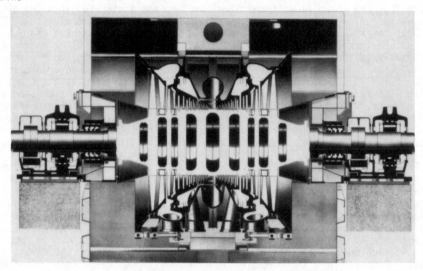

FIGURE 1.2.4 LP turbine 1800 min^{-1}

FIGURE 1.2.5 Inlet portion of LP turbine 1800 min^{-1}

Here, the steam flows from the horizontal, lateral admission pipes through
steam penetrations that are fastened to the outer casing with expansion joints,
into the inner casing. The inlet portion of the inner casing consists of two
half-scrolls (Fig. 1.2.5) which guide the steam with a minimum of flow losses to
the two blade flows.

The outer casing consists of 12 welded components which are bolted together.
Only the four components which connect to the condenser are welded together
at the site. The transverse end beams are separated in the horizontal plane of
the turbine axis, to permit assembly and removal of the rotor (Fig. 1.2.6). The
longitudinal beams parallel to the turbine axis always remain in their assembly
position.

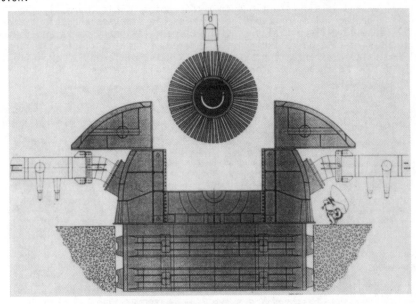

FIGURE 1.2.6 Dismantling of LP rotor 1800 min^{-1}

FIGURE 1.2.7 LP inlet casing (2 scroll halves)

The fix-point of the LP turbine is located in the center of the outer casing.
The supports on the foundation are located on the endwalls and at the corners of the outer casing in such a way that the outer casing can expand freely.
The central alignment of the outer casing is achieved by key-ways at the bearing
pedestals of their foundations.

The cast inner casing consists of the inlet scroll casing (Fig. 1.2.7) and
the subsequent blade carriers which are bolted to the scroll. Scroll and blade
carrier halves are bolted together at the horizontal plane (Fig. 1.2.8). The
inner casing is supported on each side by two cast-on brackets at the height of
the joint face. The fix-point of the inner casing, relative to the outer casing,
is located at the plane of one of the pairs of brackets, so that the inner casing
can freely expand in all directions from this point. This thermally flexible
design is a prerequisite for avoiding inadmissible stresses and deformations. Any
thermally induced differential expansion between stationary parts and the rotor
must at the same time be reliably precalculated, so that sufficiently large
radial and axial clearances can be provided. Axial differential expansions in
half-speed 1300 MW turbogenerators may attain maximum values of 30 to 40 mm.

FIGURE 1.2.8 LP casing 1800 min^{-1}

The rotor shown in figure 1.2.4 has a length of about 13 m between the two
coupling flanges and a rotor diameter of 2.64 m. Its weight (including the blading) slightly exceeds 200 tons. It is made up of forged discs and shaft ends
which are welded together. In such large half-speed rotors, this design produces
lower stress than on rotors with a central shaft and shrunk-on discs (Fig. 1.2.9).
This in turn permits the use of less highly annealed shaft steels, so that the
risk of stress corrosion cracking is also lowered (Figs. 1.2.10, 1.2.11).

The half-speed LP turbine has virtually the same blading as the full-speed
turbine, even though its dimensions are twice as large. The 2 last-stage blade
rows again have free-standing, tapered and highly twisted blades manufactured
from precision-forgings. The last stationary blade row consists of hollow profiles welded into cast half-rings.

The gland seals are bolted to the bearing pedestals and are connected to
the outer casing via flexible elements. Outer casing deformations do not affect
the clearances of the gland seals.

1.2.3 Bearings and Supports

The availability of the turbogenerator is greatly affected by rotor-dynamic
properties. A rotordynamically sound design will be less sensitive to imbalance,
so that, for example a relatively large amount of settling of the foundation and
bearing pedestals can be tolerated.

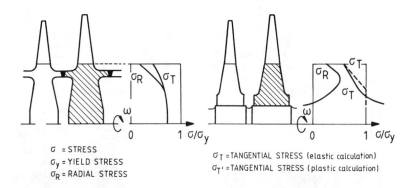

σ = STRESS
σ_y = YIELD STRESS
σ_R = RADIAL STRESS

σ_T = TANGENTIAL STRESS (elastic calculation)
$\sigma_{T'}$ = TANGENTIAL STRESS (plastic calculation)

FIGURE 1.2.9 Stresses in the welded rotor and in a rotor with shrunk-on discs

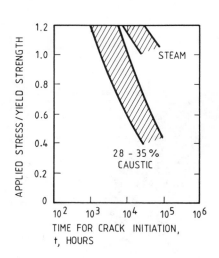

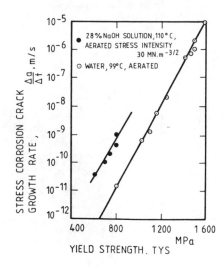

FIGURE 1.2.10 Time elapsed until crack formation in stress

FIGURE 1.2.11 Growth rate of cracking stress corrosion (cracking)

The complete rotordynamic investigation of the turbogenerator includes the rotor, bearing pedestals and foundation as shown in figure 1.2.12. The stiffness and damping of the bearings and the foundation supports must be known. Calculation methods are now being introduced which also include the dynamic properties of the foundation.

Today, fully enclosed bearings (Fig. 1.2.13), or segment bearings (Fig. 1.2.14) are used for large turbogenerators. There are three basic foundation types as shown in figure 1.2.15, in general, the spring foundations (Fig. 1.2.16) are recommended.

A. Roeder

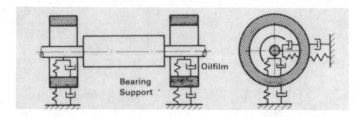

FIGURE 1.2.12 Calculation model for rotor-dynamic investigations

FIGURE 1.2.13 Radial bearing

FIGURE 1.2.14 Radial bearing with pivoted pads

Monolitic concrete foundation   Spring foundation on transoms   Spring foundation on single supports

FIGURE 1.2.15   Foundations for turbogenerators

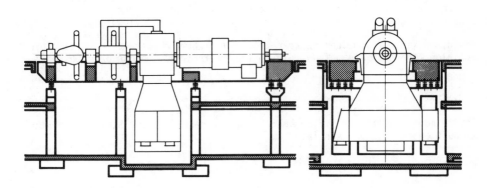

FIGURE 1.2.16   Turbogroup on low-tuned spring-supported foundations

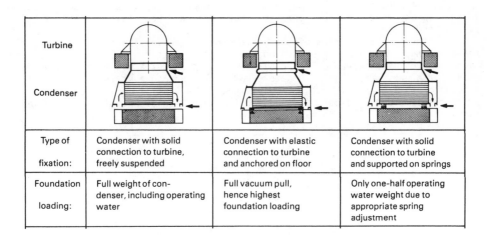

Turbine Condenser			
Type of fixation:	Condenser with solid connection to turbine, freely suspended	Condenser with elastic connection to turbine and anchored on floor	Condenser with solid connection to turbine and supported on springs
Foundation loading:	Full weight of condenser, including operating water	Full vacuum pull, hence highest foundation loading	Only one-half operating water weight due to appropriate spring adjustment

FIGURE 1.2.17   Foundation loads for various condenser fixations

1.2.4 Condenser Arrangement

The LP turbine design is essentially governed by the arrangement of the condenser. The load exerted by the condenser on the turbine foundation is highly dependent on the choice of connection between the turbine and condenser, various modern arrangements being shown in figure 1.2.17.

A condenser which is solidly connected to the turbine and is supported by springs on the machine room floor is the most favourable solution, with respect to the foundation load, and this arrangement is frequently employed.

1.3 THE LP TURBINE DESIGN

The various manufacturers of large steam turbines have standardized ranges of LP turbines. Standardization was adopted at an early date as a result of extremely high cost of developing new LP turbines. For orders received, only limited adjustments to the standard can be justified. Figure 1.3.1 presents some of the main data of a 3000 RPM LP design range. The individual type can be characterized by its exhaust area S_A. If steam flow $\dot{m}$ and condenser pressure are given, specific volume can be read from the Mollier chart and, by a simple calculation, the mean steam flow rate in the axial direction can be obtained :

$$C_{ZA} = \frac{\dot{m}v}{S_A} \qquad\qquad\qquad (1.3.1)$$

If, in addition, the mean flow angle in the exhaust annulus S_A is known, the exhaust energy can be calculated :

$$K_A = \frac{C_A^2}{2} \qquad\qquad\qquad (1.3.2)$$

This energy is assumed to be almost completely lost. The exhaust volume flow $(\dot{m} \cdot v)$ is fixed approximately by the turbine power output and the condenser conditions. Hence for a given volume flow, this lost kinetic energy will depend only upon exhaust area S_A. It becomes a matter of economic optimization to select the optimum S_A and therefore the approximate LP turbine type. This process is known as cold-end optimization.

For the standardization of the LP turbines, a design inlet pressure or at least a range of LP inlet pressure must be chosen. Figures 1.3.2 and 1.3.3 show heat balances of a modern 600 MW coal-fired plant and of a 900 MW nuclear power plant for a pressurized water reactor (PWR). The pressure at the LP inlet of saturated steam plants is considerably higher than that of conventional coal power plants. Cold-end optimization and the selection of LP inlet pressure are considered in the following paragraphs.

Speed min−1	LP-Type	Mean diameter of last stage mm	Length of last stage mm	Tip speed m/s	Exhaust annulus area m²
3000	D 48	2100	600	424	4,0
	D 50	2335	665	471	4,89
	D 52	2376	792	498	5,9
	D 54	2601	867	545	7,1
	D 56	2850	950	596	8,5

FIGURE 1.3.1 LP turbine types for 3000 min^{-1}

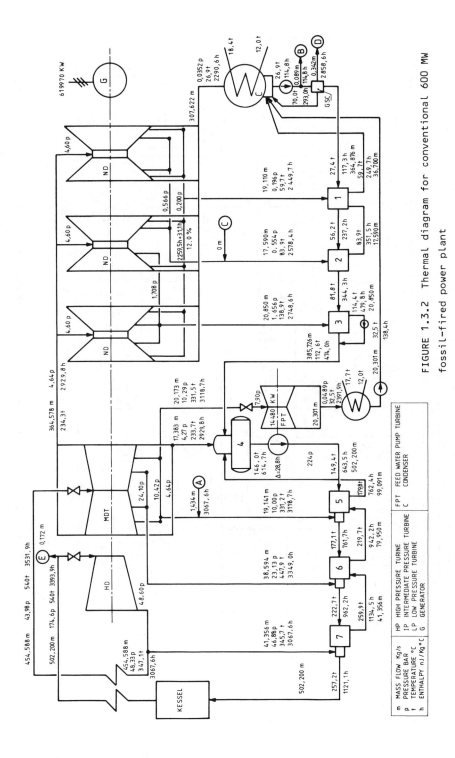

FIGURE 1.3.2 Thermal diagram for conventional 600 MW fossil-fired power plant

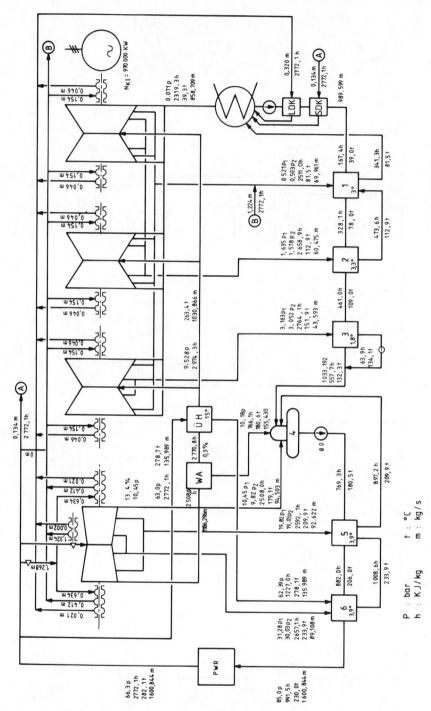

FIGURE 1.3.3 Thermal diagram for a 900 MW nuclear power plant

P : bar   t : °C

h : KJ/kg   m : kg/s

12

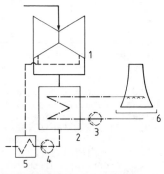

1 LP-Turbine   4 Extraction pump
2 Condenser   5 LP Feedheater 1
3 CW Pump   6 Cooling Tower

a)   Main components

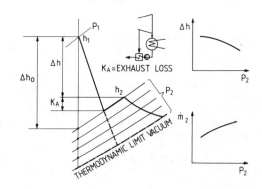

b)   Last stage expansion

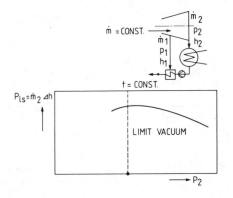

c)   Thermodynamics

d)   Input data

Rated output	MW	330
Wet bulb temperature	°C	3
Rel. lummidity	%	78
Price per caloric unit	US$/Gcal	6,5
Spec. price of condenser	US$/m²	100
Cost of cooling water supply	US$/m³s⁻¹	80·10³
Make up water cost	US$/m³	0,01
Period of operations		20 Years
		5000 h/a
Interest rate		10 %

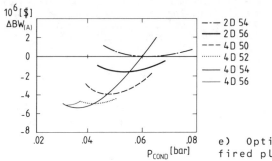

e)   Optimum curves 330 MW fossil-fired plants

FIGURE 1.3.4   Optimization of cold end

1.3.1 Cold-End Optimization

For cold-end optimization, the LP turbine, condenser, lowest LP feedheater, and
the cooling water system are taken into consideration (Fig. 1.3.4-a). For this
optimization, one cannot freely choose the condenser pressure or "vacuum".
Climatic conditions, such as air temperature, humidity, and cooling water temperature, may limit the choice of the vacuum. But within the given limits, the investigation will provide the economic optimum. In addition to the above mentioned
plant components, all the economic parameters, e.g. the capitalized operating
costs, must also be taken into account. The investigation will provide the
optimum operating parameters, such as the cooling water flow and the temperature
at the condenser outlet.

Figure 1.3.4-b shows the expansion line of the steam in the last LP turbine
stage on the "Mollier chart". The enthalpy difference Δh corresponds to the
specific work of this stage; K_A is the exhaust loss previously defined by
equation (1.3.2).

Δh will rise with decreasing condenser pressure, as shown in figure 1.3.4-b,
but at the same time, the temperature of the condensate at the inlet to the lowest LP feedwater heater falls. The extraction flow rate $\dot{m}_1$ (Fig. 1.3.4-c) will
increase and the steam flow $\dot{m}_2$, through the end stage, will decrease accordingly.
The output from the last stage is thus :

$$P_{LS} = \dot{m}_2 \cdot \Delta h$$

K_A increases with decreasing p_2 and, at a certain p_2, the choking point of the
final stage is reached. Any further decrease of p_2 increases exhaust losses,
Δh remaining unchanged. The last stage output will therefore follow the trend
shown in figure 1.3.4-c.

Each selected condenser pressure will therefore determine a set of parameters
including turbine output, cost of components and capitalized operating costs.
For given climatic conditions, the condenser pressure can be lowered only by
taking measures which increase the cost of the plant and the cooling system, for
example by increasing cooling water flow and condenser heat transfer surface.
Similarly the added thermal resistance of dry or wet cooling towers will generally
require increased condenser size and result in a higher optimum condenser pressure.

The results of this cold-end optimization can be demonstrated by considering
the example of a conventional 330 MW power plant. Figure 1.3.4-d presents the
basic data required for the investigation; figure 1.3.4-e summarizes the results.
Here, the costs in comparison with a reference plant are plotted against the condenser pressure, for various 2 and 4 flow LP turbines. A specific optimum can be
found for each LP turbine with, in each case, the optimum combination of the components shown in figure 1.3.4-a. For this example, the overall optimum corresponds
to a 4-flow LP turbine of the D54 type with a condenser pressure of 0.04 bar.

1.3.2 LP Inlet Pressure

1.3.2.1 Fossil-Fired Power Plant

Turbines for fossil-fired power plants, as a rule, incorporate at least twice as
many flows in the LP turbine as in the IP turbine, due to the enormous increase
in the steam volume between the IP turbine and the condenser. Thus, the 600 MW
turbine for the heat balance presented in figure 1.3.2 possesses only 1 double-
flow IP, but 4 LP flows (Fig. 1.3.5).

The efficiency of the turbine stages can be presented as a function of the
flow coefficient δ :

$$\delta = \frac{\dot{V}}{r^2 \cdot u} \tag{1.3.3}$$

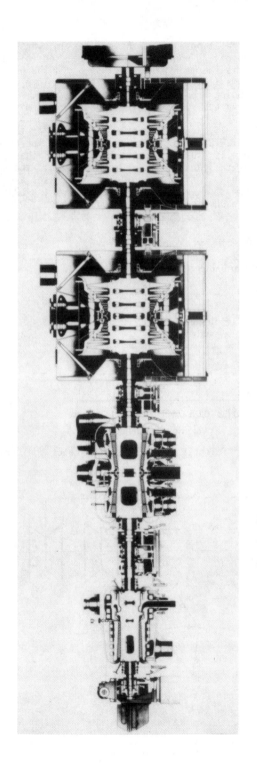

FIGURE 1.3.5 600 MW Turbogroup 3000 min^{-1}

15

Figure 1.3.6 shows this relationship for the two design types, the impulse turbine and the reaction turbine. In equation (1.3.3) $\dot{V}$ is the volume flow of the stage under consideration, r is the mean blade radius, and u the mean circumferential speed. To achieve maximum stage efficiencies, a volume coefficient larger than 0.5 is sought. Going from the IP to the LP turbine, $\dot{V}$ is reduced to half or less of its previous value, since the IP volume-flow is divided between at least two LP flows.

In addition, as a rule, r is larger at the LP inlet than at the IP outlet. The denominator of equation (1.3.3) is proportional to r^3 and increases accordingly. At the LP inlet, δ thus is usually considerably smaller than at the IP outlet. Figure 1.3.6 shows that the larger δ is at the IP outlet, the smaller is its effect upon the stage efficiency. At the IP outlet, δ increases as the pressure at this point decreases. From this we can conclude that for the LP inlet in conventional power plants it is economically advantageous to have low pressures.

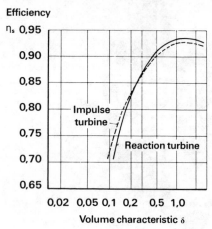

FIGURE 1.3.6   Stage efficiencies of impulse and reaction turbines

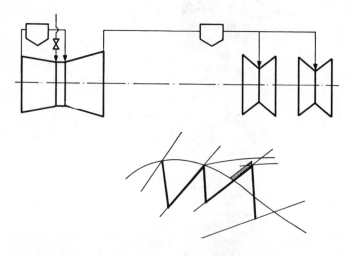

FIGURE 1.3.7   Modern design of a saturated steam plant
with high speed separator-reheater

Today, the standard design usually provides the seven feedheaters of which
the fourth and last LP feedheater receives its extraction steam for the IP outlet.
The thermodynamically optimal layout of the feedheating system furnishes a further
criteria for the LP inlet pressure. The heating of the feed water must be sub-
divided in an optimal manner. By choosing the IP outlet as the location of the
fourth extraction, the pressure at this extraction is also fixed.

1.3.2.2 *Nuclear Power Plant*

Basically, the same rules apply to the turbines of the nuclear power plant as for
the conventional plant, but it should be noted that the volume flows and hence
also the volume coefficient δ are considerably larger in saturated steam turbines
than in conventional turbines. In nuclear power plants the overall plant costs
are greatly affected by variations of the LP inlet pressure. In a thorough investigation of the optimum LP inlet pressure, the costs of the moisture separator-
reheater, the condenser, the cross-over pipes, the safety valves between the MSR
and the LP turbines, and the turbine house must also be taken into account.

With decreasing LP inlet pressure, the MSR dimensions and their costs increase rapidly, along with the spatial requirements of the turbine house. This
shifts the optimum LP inlet pressure of the saturated steam plant to values of
between 10 and 14 bar.

The large dimensions of present day MSR's are due to the relatively small flow
velocities still prescribed for this piece of equipment. Smaller, high-speed
separators are being developed. In future saturated steam plants, the optimum
LP inlet pressures will presumably be pushed again toward smaller values, thus
permitting better turbine efficiencies. Figure 1.3.7 shows a possible design.

1.4 SPECIAL DESIGN PROBLEMS OF THE LP TURBINES

Following the discussion of the design process in the preceding paragraphs and
economic optimization of the steam conditions at the inlet and outlet of the
LP turbines, some of the design problems pertaining specifically to the large
LP turbine will now be described.

1.4.1 Inlet Casing

The inlet casing of the LP turbine generally consists of an annular space into
which the cross-over pipes are connected (Fig. 1.4.1). The steam leaves this
annular space in the radial inward direction and reaches the inlet to the blading
(Fig. 1.4.2) after a 90° deflection towards the axial direction.

From the cross-over pipe to the blading inlet, the steam thus suffers sudden
cross-sectional and directional changes. Furthermore, partial flows meet from
opposite directions and the steam is admitted to the blading in a non-uniform
manner. This type of inlet casing thus entails relatively high flow losses which
can only be limited to an acceptable value, if the flow rates remain relatively
small.

The requirement to maintain small flow velocities forces the designer to
contemplate large dimensions which, in turn, cause larger stresses and deformations, due to pressure and temperature differences.

Some large modern LP turbines employ an inlet casing consisting of 2 scroll
halves, shown earlier in figure 1.2.7. Such scrolls are familiar from water turbine designs. They avoid the typical flow losses of the customary torus casing
by guiding the steam smoothly to the inlet of the first moving rows (Fig. 1.4.3),
thus permitting considerably greater flow rates, along with smaller dimensions.
The more compact inlet casing suffers less stress and deformation.

A. Roeder

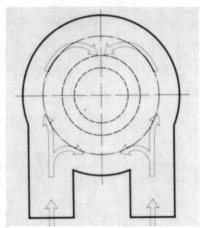

FIGURE 1.4.1 Conventional inlet casing (Torus)

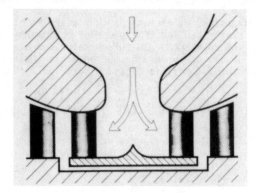

FIGURE 1.4.2 LP-1 stage, axial design

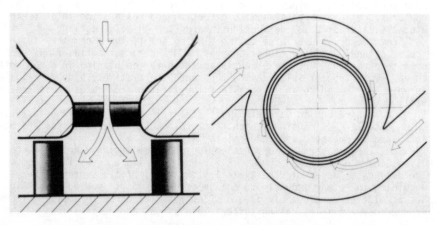

FIGURE 1.4.3 LP-1 stage, radial-axial design

Before introducing such a new concept, the positioning of the cross-over pipes must be considered and the fluid mechanics engineer has to provide calculation procedures for the dimensioning of this new casing. As a rule, measurements are also made in a special test plant, to confirm theoretical predictions.

In the present case, the actual design problem consisted of the development of suitable computational procedures for the layout of the scroll and, in particular, for the transition from the pipe to the scroll flow.

1.4.2 LP Rotor and Blading

The enthalpy drop in the LP blading is dictated by the given steam data at the inlet and outlet.

When choosing the individual stage gradients, the extraction locations of the feedheater system must be taken into account (Fig. 1.4.4). The enthalpy drop

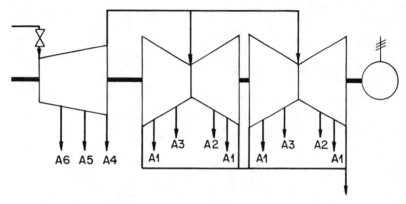

FIGURE 1.4.4 LP extractions

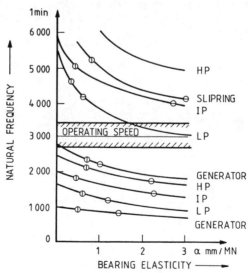

⊕ VERTICAL FREQUENCY
⊖ HORIZONTAL FREQUENCY

FIGURE 1.4.5 Critical speeds of a turbogenerator

Δh of the various stages can be calculated from the work coefficient μ and the average circumferential speed u

$$\Delta h = \mu \cdot u^2 \qquad\qquad\qquad\qquad (1.4.1)$$

For reaction stages work coefficients μ of between 1 and 1,2 yield optimum efficiencies. For impulse stages higher μ values must be chosen. If all stages were to be designed for optimum work coefficient μ, whilst simultaneously satisfying extraction requirements, the circumferential speed u would be a prescribed. The mean blade radii, however, can only be fixed by taking the requirements of the rotor design into account. Thus the rotor and the blading must be studied together when determining the optimum design.

The costs of the bladed rotor amount to about 40% of the total LP turbine costs. The cost of the rotor depends to a large extent upon the hub diameter. To keep the total costs to a minimum, the hub diameter should therefore be kept as small as possible.

If the stages are optimally designed (with a fixed μ), the heat drop per stage will also decrease with decreasing hub diameter, while the number of stages and hence the LP rotor length will increase. This thinner, longer rotor may lead to difficult rotor-dynamics and may result in an inadmissible frequency range near the intended operating speed (Fig. 1.4.5).

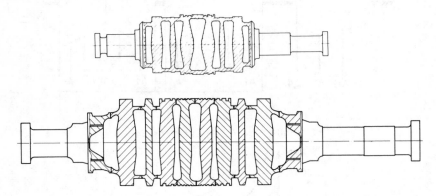

FIGURE 1.4.6 Full speed and half-speed LP rotors of welded design

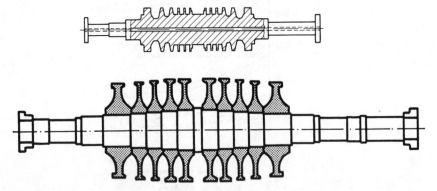

FIGURE 1.4.7. Full-speed and half-speed LP-rotors,
monoblock rotor and shrink-on disc design

Stage number and hub diameter are the main parameters which determine the construction and the design of the LP rotor, as shown in figures 1.4.6 and 1.4.7. Thus, the welded half-speed LP rotors are made up of a greater number of forgings, in order to reduce the weight. Many turbine manufacturers still design LP rotors with a number of shrunk-on discs (corresponding to the number of stages). Of course, each alternative has its own particular stress conditions in the rotor and these stresses greatly affect the operating safety.

Each LP turbine type can be characterized by its outlet area S_A,

$$S_A = \pi D_M^2 \left(\frac{H}{D_M}\right) \qquad (1.4.2)$$

If during the economic optimization process the hub diameter and thus the mean blade diameter (Fig. 1.4.8) is reduced, the blade length/diameter ratio H/D_M must be increased. H/D_M is, however, a parameter which greatly influences the three-dimensional flow in the last stage. Today's customary values lie between 0.3 and 0.35. Extreme cases may reach 0.38. The larger the H/D_M, the more difficult the sound aerodynamical design of the blade becomes.

An increase in the ratio H/D_M for a fixed S_A leads to longer last stage blade (LSB). The risk of blade vibration increases with increasing length, since, as natural frequencies decrease, the probability of exciting a vibrational mode will increase. In LP turbines of 3000 RPM, free-standing moving blades up to 1050 mm in length, and tied moving blades up to 1044 mm are in service (Fig. 1.4.9). These blades are forged from 12 to 13% chromium steel; to date titanium has only been used in exceptional cases.

The LP turbine layout is very dependent upon the effort spent in the development of an operationally safe LSB. Efficient computational procedures are now available for three dimensional flow calculations (Fig. 1.4.10). These programs assume, however, that the flow angles can be calculated for any blade cascades on conical stream surfaces and also for transonic flow. For this task, three-dimensional time marching procedures are being developed. Flow losses in LSB cascades can be predicted only partially because there is a lack of reliable boundary layer procedures for cascades with shocks and separations.

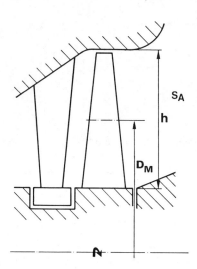

FIGURE 1.4.8 Main dimensions of the LP last stage (LB)

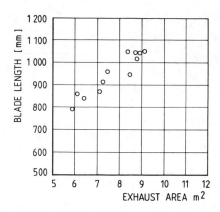

FIGURE 1.4.9 Lengths of last stage blades in service

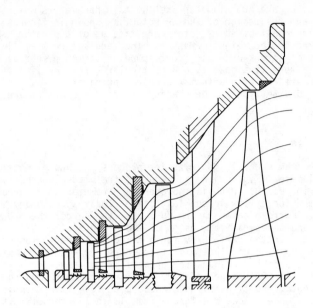

FIGURE 1.4.10   Meridian streamlines of an LP turbine

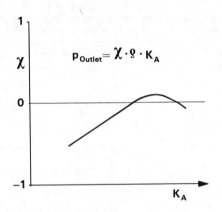

FIGURE 1.4.11   Pressure losses in the LP outlet

Finite element methods are available for the design of LP rotors, for the reliable calculation of stationary and transient stresses. Local stress peaks can be avoided and the material utilization improved, without decreasing the high safety factors.

Rotor-dynamic properties can be accurately determined by means of complex procedures for critical speeds, out-of-balance vibrations and stability. In the field of fluid mechanics, there are as yet no procedures which permit a detailed loss analysis of the wet steam flow through the LP blading to be made. Thus, a gradual improvement of the LP turbine fluid mechanic design can only be achieved through expensive measurements in test installations. Hence the economic optimization of LP turbine design cannot yet contain all design parameters.

Further examples of partially solved problems are : The extreme increase in volume along the flow path in the LP blading leads to highly conical channels (Fig. 1.4.10). The flow cross section between the various blade rows increases in some cases very considerably and this may cause separation at the walls. Highly conical cascade flow affects the outlet angle and must be taken into account in the profile development.

1.4.3 LP Outlet

In the cold-end optimization, the simplifying assumption was made that the outlet energy K_A is lost. Measurements at the LP outlet, however, show that in a favourably arranged diffusor and casing some of this exhaust energy can be regained at certain operating points. In other cases, however, the exhaust loss may be even higher. The proper cold-end optimization requires that, for each LP turbine type, the exhaust loss characteristic is known, e.g. in the form of

$$\chi = f\ (K_A) \qquad\qquad\qquad (1.4.3)$$

and χ, in accordance with figure 1.4.11 is known.

Generally, applicable theories for exhaust system designs are not yet available. So, usually, for new designs approximate methods must be used.

1.5 CONCLUSIONS

The design problem of the LP turbine, today, is characterized by the requirement of finding economically optimum solutions. This will only be successful if the turbine efficiencies and the total costs of all the alternatives studied are evaluated according to economic principles. The proven design element of today's operating LP turbines represent the basis for every further development. Such developmental steps must always be guided by criteria of cost and benefit. For LP turbines, this quantification of the benefits, resulting from the modification of certain design details, can at present be only partially achieved.

Design methods are very successful, however, in the mechanical area, because accurate computational methods for stress calculations and the casting material of manufacturing are available. In the fluid mechanics area on the other hand suitable methods for the reliable prediction of benefits do not always exist.

NOMENCLATURE

		Subscripts	
C	velocity	A	outlet
D	diameter	M	mean diameter
h	enthalpy	LS	last stage
H	blade height	z	axial
K	kinetic energy		
$\dot{m}$	mass flow rate		
p	pressure		
P	power		
r	radius		
S	surface		
u	circumferential velocity		
v	specific volume		
$\dot{V}$	volume flow rate		
δ	flow coefficient		
μ	loading coefficient		
δ	stress		

CHAPTER 2

Throughflow Design Methods

2.1 STREAMLINE CURVATURE METHOD

M. Troilo

2.1.1 Foreword

Through Flow (TF) calculations for radial and axial turbomachines have been in use for about twenty years, that is, since fast (electronic) computers have been available.

Computer codes, during this time, have seen increasing development, becoming more and more complete and reliable, especially with respect to the inclusion of loss and flow angle deviation correlations, the extension to multi-stage machines, the off-design performance prediction and the treatment of special problems, such as that posed by the two phase flow in steam turbines.

The pertinent scientific and technical literature includes a large number of papers and reports covering the theoretical basis of the method, whilst a few are concerned with the problems arising in the computer code formulation. Many of these problems are indeed not trivial, and are connected with physical aspects of the flow.

2.1.2 Brief Historical Survey

The work of C.H. Wu [2.1,2.2] is generally considered the first complete theoretical analysis of the flow through a turbomachine; actually the first historical paper on the subject is attributed to H. Lorenz (1905), who is cited by Wu himself [2.3].

The highly complex problem of the three-dimensional flow calculation was first simplified by separating the flow into two distinct surfaces :
a)   the blade-to-blade flow, i.e., the flow on Wu's S_1-type surfaces;
b)   the meridional flow, i.e., the flow on Wu's S_2- type surfaces.
The calculations of these flows can eventually be coupled together to obtain improved solutions.
Further simplifications concern the shape of the S_1-S_2 surfaces to be considered.

The scope of the present lecture is restricted to the problem of meridional through-flow calculation in axial flow (steam) turbines. For the meridional flow the fundamental equation is then the so-called Radial Equilibrium Equation (REE), i.e., the projection of the motion equation along a radial direction. The well known paper of L.H. Smith [2.4] on this subject should be mentioned as very important for two main reasons :
a)   it gives a rigorous formulation of REE specifying the conditions under which the axisymmetric equation can be applied to circumferential-averaged quantities;
b)   it reveals that the REE has been applied for industrial calculation purposes since 1959.

A more comprehensive analysis of the TF equations, including a detailed formulation of the remainder terms appearing when the circumferential-averaging is
applied, was included by Chauvin in his course on turbomachinery fluid dynamics
at VKI [2.5].
    As far as the numerical solution of TF equations is concerned, it is easy to
recognize that the main impetus came from specialists involved in industrial
activities. Indeed the numerical solution poses a number of problems and setting-
up proper methods and procedures requires an amount of work that is difficult to
justify in an academic context.
    The product of the cooperative work of scientists and industrial engineers has
given rise to a well-defined field of research, that may be called "Calculation
Methods in Turbomachinery Fluid Dynamics".
    In this field, as far as the meridional flow is concerned, three main methods
have been applied :
a) the SCM (streamline curvature method), attributed to Novak [2.6,2.7] ;
b) the matrix method, attributed to Marsh [2.8] ;
c) the finite element method, applied to this problem by Hirsch [2.9].
    An interesting comparison between the SCM and the matrix method can be found
in [2.10] ; the SCM has been generally preferred for industrial purpose calculations.
    In the last decade, three kinds of contributions to the subject can be listed:
a) papers concerning more specifically the TF calculation procedures; in this
group may be counted, for example, the reports of Renaudin & Somm [2.11], Cox
[2.12,2.13], Denton [2.14], Novak [2.15], Doria & Troilo [2.16] ;
b) papers that contribute by giving the mathematical formulation a more and
more rigorous basis : Veuillot [2.17], Thiaville [2.18], Horlock [2.19].
c) papers considering limiting situations and applicability criteria such as
those by Fruehauf [2.20] and Schröder & Schüster [2.21].
Finally, the state-of-the-art of the meridional TF calculations in axial turbomachines has been established in [2.22] by means of many contributions collected
and edited by Denton and Hirsch.

2.1.3   Synthesis of Mathematical Formulation

This lecture deals with the SCM which is the method most used for industrial
applications. It consists essentially of a set of one-dimensional calculations,
along meridional streamlines, coupled together by REE. Along the meridional
streamlines the energy equation in the usual enthalpy formulation is used, together with kinematic relationships to switch from the absolute to the relative

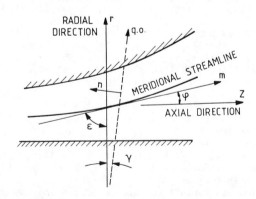

FIGURE 2.1.1 Meridional stream channel, showing a meridional streamline,
a quasi-orthogonal and angle definitions for equations (2.1.2,2.1.3,2.1.9)

flow and vice-versa, and geometric relationships between flow angles and velocity components. The radial distribution of quantities is calculated by integration of REE, this latter equation being used in different forms.

The REE in the form given by Smith [2.4], for a region where no body forces nor blockage effects exist, is as follows (for notations see list of symbols and Fig. 2.1.1) :

$$\frac{1}{\rho}\left(\frac{\partial p}{\partial r}\right)_z = \frac{1-M_z^2}{1-M_m^2}\left[\frac{C_\theta^2}{r} - \left(\frac{d^2r}{dz^2}\right)_\psi C_z^2\right] + \frac{C_r}{1-M_m^2}\cdot\frac{C_z}{r}\left(\frac{\partial(r\sigma)}{\partial r}\right)_z \tag{2.1.1}$$

This is a complete REE for steady axisymmetric, inviscid, compressible flow. As pointed out by Smith [2.4], the terms appearing at the RHS of equation (2.1.1) account for radial accelerations due to various effects :

$\frac{1-M_z^2}{1-M_m^2}\cdot\frac{C_\theta^2}{r}$ is the radial acceleration due to the main swirl, obtained by the pure centripetal field C_θ^2/r, magnified, if the meridional flow is subsonic ($M_m < 1$) by a factor $(1-M_z^2)/(1-M_m^2)$, which is greater than unity for $\sigma \neq 0$. This factor accounts for a radial component of streamwise pressure gradient due to compressibility effects when the velocity is changing along streamlines;

$\frac{1-M_z^2}{1-M_m^2}\cdot\left(\frac{d^2r}{dz^2}\right)_\psi C_z^2$ is the radial acceleration due to the curvature of the meridional streamlines, obtained as the sum of a pure centripetal field component, given by $-\cos^2\varphi(d^2r/d_z^2)_\psi C_z^2$, and a radial component of the streamwise pressure gradient due to the streamtube area variations imposed by the curvature itself, given by $-(d^2r/dz^2)_\psi C_z^2\ (\sin^2\varphi/(1-M_m^2))$

$\frac{C_r}{1-M_m^2}\frac{C_z}{r}\left(\frac{\partial(r\sigma)}{\partial r}\right)_z$ is the radial acceleration due to the slope gradient of the meridional streamlines, given by the radial component of the streamwise pressure gradient due to streamtube area variations resulting from the gradient itself.

It is interesting to compare equation (2.1.1) with that written under the same hypothesis by other authors, e.g. Denton [2.14] :

$$\frac{1}{2}\left(\frac{\partial(C_m^2)}{\partial r}\right)_z = \left(\frac{\partial h_t}{\partial r}\right)_z - T\left(\frac{\partial s}{\partial r}\right)_z - \frac{1}{2r^2}\left(\frac{\partial(r^2 C_\theta^2)}{\partial r}\right)_z + \frac{C_m^2}{r_c}\sin\varepsilon + C_m\frac{dC_m}{dm}\cos\varepsilon \tag{2.1.2}$$

practically the same as that by Novak [2.15] :

$$\frac{1}{2}\left(\frac{\partial(C_m^2)}{\partial r}\right)_z = \left(\frac{\partial h_t}{\partial r}\right)_z - h\left(\frac{\partial(s/c_p)}{\partial r}\right)_z - \frac{C_\theta}{r}\left(\frac{\partial(r C_\theta)}{\partial r}\right)_z + C_m\frac{dC_m}{dm}\sin\varphi + \frac{C_m^2}{r_c}\cos\varphi \tag{2.1.3}$$

which gives a further detail as :

$$C_m\frac{dC_m}{dm} = \frac{C_m^2}{1-M_m^2}\left[\frac{\sigma}{r_c} - \frac{1}{\sin^2\varphi}\left(\frac{\partial\varphi}{\partial r}\right)_z - (1-M_\theta^2)\frac{\sin\varphi}{r}\right] \tag{2.1.4}$$

A quite different formulation may be obtained if the stream function ψ defined as

$$\left(\frac{d\psi}{dr}\right)_z = 2\pi\rho C_z r dr \tag{2.1.5}$$

is substituted; this has been developed to useful detail by Cox [2.12,2.13] giving:

$$\left(\frac{\partial p}{\partial \psi}\right)_z = \frac{1}{2\pi r}\left[\frac{C_\theta^2}{C_z r} - C_z\left(\frac{\partial \sigma}{\partial z}\right)_\psi - \sigma\left(\frac{\partial C_z}{\partial z}\right)_\psi\right] \tag{2.1.6}$$

which may be considered a very compact formulation if the term $\partial C_z/\partial z$ is numerically evaluated as suggested in [2.13].

The most compact form is undoubtedly that used by Renaudin & Somm [2.11]

$$\frac{1}{\rho}\frac{\partial p}{\partial r} = \frac{C_\theta^2}{r} - \frac{dC_r}{dt} \tag{2.1.7}$$

observing that the term dC_r/dt has to be evaluated numerically, in a direct, even if rather complex way. In this form (2.1.7), the physical basis of the REE can be seen directly, the radial pressure gradient being created by (i) the radial (centripetal) acceleration due to the swirl, and (ii) the radial acceleration due to the variation of the radial component of the velocity, which is given by the substantial derivative of C_r with respect to time.

Observing that $C_z = dz/dt$, is the derivative taken along a streamline (ψ = const), it is easy to recognize that $dC_r/dt = C_z(dC_r/dz)_\psi$, and remembering that $C_r = \sigma C_z$, the form (2.1.6) is quickly obtained from (2.1.7) without recourse to the original complex mathematics (see [2.12]).

With minor changes, all the above forms of REE are written for directions, tilted with respect to the radial one, that have been called q.o. (quasi-orthogonals). This need arises from two considerations :
a) in a multistage machine, calculation stations will be positioned along the blade edges (leading or trailing) as in figure 2.1.2 which are not generally radial :

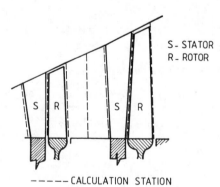

S - STATOR
R - ROTOR

----- CALCULATION STATION

FIGURE 2.1.2 Calculation stations in the meridional plane of a steam turbine

b) if the chosen directions were truly orthogonals to the streamlines, the terms defined in equation (2.1.4) disappear, as observed e.g. in [2.15]. Equation (2.1.3) along a q.o. may be written as

$$\frac{1}{2}\frac{\partial C_m^2}{\partial q} = \frac{\partial h_t}{\partial q} - h\frac{\partial(s/c_p)}{\partial q} - \frac{C_\theta}{r}\frac{\partial(rC_\theta)}{\partial q} + C_m\frac{\partial C_m}{\partial m}\sin(\gamma+\varphi) + \frac{C_m^2}{r_c}\cos(\gamma+\varphi) \qquad (2.1.8)$$

and likewise equation (2.1.6) is written :

$$\left(\frac{\partial p}{\partial \psi}\right)_n = \frac{1}{2\pi r(1-\sigma\tan\gamma)}\left[\frac{C_\theta^2}{C_z r} - C_z\left(\frac{\partial\sigma}{\partial z}\right)_\psi - (\sigma+\tan\gamma)\left(\frac{\partial C_z}{\partial z}\right)_\psi\right] \qquad (2.1.9)$$

The equation above has been chosen for a computer code which has proven useful for industrial design purposes.

The other equations necessary do not pose particular problems. They are :
- energy equation in ducts and fixed blades :

$$h_t(\psi)_{\text{previous station}} = h(\psi) + \frac{C^2(\psi)}{2} \qquad (2.1.10)$$

- energy equation in moving blades :

$$h_{t,R}(\psi)_{\text{previous station}} = h(\psi) + \frac{W^2(\psi)}{2} - \frac{\Omega^2 r^2}{2} \qquad (2.1.11)$$

To correctly interpret the meaning of "previous stations" it is intended, of course, that at least one calculation station exists for any blade row. Moreover, it can be observed that the validity of the energy equation along streamlines is a decisive argument to choose the mathematical formulation in terms of the stream function.
- continuity equation :

$$m = \int_{r_H}^{r_T} 2\pi r\rho C_z(1-\sigma\tan\gamma)dr \qquad (2.1.12)$$

- kinematic and geometrical relationships :

$$C = \sqrt{C_\theta^2+C_z^2+C_r^2} \qquad (2.1.13)$$

$$W = \sqrt{W_\theta^2+W_z^2+W_r^2} \qquad (2.1.14)$$

$$\sigma = C_r / C_z \qquad (2.1.15)$$

$$C_z = C_\theta\tan\alpha \qquad (2.1.16)$$

$$C_z = W_\theta\tan\beta \qquad (2.1.17)$$

$$C_\theta = W_\theta+\Omega r \qquad (2.1.18)$$

- state equations :

$$f(p,h,s) = 0 \qquad\qquad (2.1.19)$$

$$f(p,h,p) = 0 \qquad\qquad (2.1.20)$$

- expansion line equation :

$$f(p,h) = 0 \qquad\qquad (2.1.21)$$

The expansion line is in fact defined by the magnitude of the losses, so it will be redefined each time the losses are updated during the iteration process.

2.1.4   Boundary Conditions

For the TF calculation of large steam turbines, and especially their low pressure sections, boundary conditions have been chosen remembering that :
a)   the discharge pressure is defined by the condenser whose high thermal capacity ensures a constant pressure operation;
b)   as a rule the flow in the last stage is transonic. Then, condition (a) can only be met by adapting the inlet total pressure.
Boundary conditions must therefore always include the static discharge pressure as fixed, on at least one point on the last downstream calculation station. So there are two possibilities :
a)   the static discharge pressure and the inlet total conditions are given, and the mass flow rate has to be calculated;
b)   the static discharge conditions and the mass flow rate are given, and the inlet total conditions have to be calculated.
Both methods are fully equivalent if the characteristic curve of the machine has to be found; in the following, reference is made to case (b).

2.1.5   Numerical Integration of REE

From the calculation point of view, all formulations may be distinguished as belonging to one of two groups, depending on whether or not the energy equations were substituted in the equation of motion :
a)   when substitution takes place (e.g. [2.14,2.15])

$$\frac{dC_m^2}{dq} = f(C_m^2,q) \qquad \text{or} \qquad \frac{dC_m^2}{d\psi} = f(C_m^2,\psi) \qquad (F1)$$

b)   when it does not (e.g. [2.12,2.13,2.23])

$$\frac{dp}{dq} = f(p,q) \qquad \text{or} \qquad \frac{dp}{d\psi} = f(p,\psi) \qquad (F2)$$

The use of one of the two forms, (F1) or (F2) in computer codes leads to different overall calculation schemes, that are worth mentioning.
In the form (F1), the function f includes the term Q

$$Q = \frac{dh_t}{dq} - T \frac{ds}{dq} \qquad \text{or} \qquad Q = \frac{dh_t}{d\psi} - T \frac{ds}{d\psi}$$

which appears explicitly. In one respect this is attractive, because in many cases the term Q can be equated to zero, without significant error; on the other hand, when for instance a strongly non uniform distribution of losses exists, the term Q cannot be disregarded, and has to be calculated by numerical differentiation which is an error-amplifying process. Furthermore, this evaluation is possible only at an outer iteration level, because difference formula can be conveniently applied when the functions $h_t(q),s(q)$ or $h_t(\psi),s(\psi)$ are known in all points on the q.o.

Integration of REE in form (F2) is straightforward, and in the SCM requires only the slope and curvature terms obtained from previous iterations.

Reference is made for clarity to equation (2.1.9) which is the REE formulation adopted in the particular computer code whose results will be presented later on. Some quantity must be prescribed along the radius: if this quantity is the flow angle, then one is solving the direct problem.

In calculating flow in the annular space, the conservation of angular momentum along streamline is commonly assumed, even if this is true only for inviscid flows :

$$\left[C_\theta r(\psi) \right]_{\text{previous station}} = C_\theta r(\psi) \qquad (2.1.22)$$

Considering equation (2.1.9) as a common differential equation of the form

$$\frac{dp}{d\psi} = f(p,\psi) \qquad (2.1.23)$$

it is easy to recognize that, when equation (2.1.22) is applied, i.e., $C_\theta = C_\theta(r)$ in equation (2.1.9), equation (2.1.23) does not satisfy the Lipschitz condition on p :

$$\left[\frac{f(p,\psi)-f(p+\Delta p,\psi)}{\Delta p} \right] \leqslant G \qquad (2.1.24)$$

with G finite and $\Delta p < K$ arbitrary. This happens because for small α, with a given C_θ, C_z is positive only for those p values such that $C > \sqrt{C_\theta^2 + C_r^2}$. When C_z approaches its lowest limit, $f \to \infty$.

Hence a fundamental condition for a stable integration process is violated. Moreover, because for practical reasons the discretization cannot be pushed beyond some limits, when the function f tends to grow, the product $|f \cdot \Delta p|$ will exceed any limit dictated by stability requirements in every numerical integration method.

In the calculation of large steam turbines, the problem mentioned above has been experienced in the calculation of the flow in the annulus between the stator and rotor blades, where a small angle α is present. Because a calculation station at the inlet of the rotor blade may be preferred for a better estimate of the incident flow angles, some method must be adopted to avoid the instability. One suggestion is as follows : because the stability conditions are violated for too small values of the angle α, or, which is the same, for too steeply decreasing C_z with respect to the previous station, a lower limit can be set for C_z, e.g. defining an expected value not less than an empirical fraction of that pertaining to the previous station. In this way the function f in equation (2.1.23) has been modified, in a (p,ψ) region where the solution does not fall, to make it Lipschitzian.

Finally, some remarks are worthwhile with respect to the integration method. The particular case of the turbine flow calculation using the SCM does not allow the use of standard numerical methods such as Runge-Kutta, Milne and so on. In fact, with the SCM one deals with a fixed streamline number, and a number of terms in equation (2.1.9) are calculated by numerical techniques from the whole streamline pattern (they are the slope and curvature terms). Moreover, calculations involved in industrial design, are based on blade geometries which are given to high precision, for quality assurance aims, in selected blade sections.

A successful use of standard methods requires, even forgetting the variable integration step, the calculation of the f-function at the interior of the step,
and this information is not available in the SCM.

Two methods have been proposed by specialists in the field, and they have been
proven satisfactory. They are :
a) The modified or improved Euler's method, proposed by Katsanis [2.24]. This
is basically a predictor-corrector method with these formulas :

$$p_{j+1}^{PR} = p_j + \Delta\psi \cdot f(p_j, \psi_j) \qquad\qquad (2.1.25)$$

$$p_{j+1}^{COR} = p_j + \frac{\Delta\psi}{2} \left[f(p_j, \psi_j) + f(p_{j+1}^{PR}, \psi_{j+1}) \right] \qquad\qquad (2.1.26)$$

An iteration could be performed in the corrector formula, putting $p_{j+1}^{PR} = p_{j+1}^{COR}$
until convergence is obtained within a specified error. In the formulas (2.1.25,
2.1.26) the index j refers to the current streamline.
b) The Picard's method, suggested by Cox [2.12, 2.13]. This method is based on
the estimate $p^{OLD}(\psi)$ of the solution, that permits an improved one to be calculated as :

$$p^{NEW}(\psi) = \int_0^\psi f(p^{OLD}, \psi) d\psi \qquad\qquad (2.1.27)$$

followed eventually by iterations putting $p^{OLD} = p^{NEW}$.
Good results are obtained using the method (a) in the early outer iterations,
and the faster method (b) near the global convergence, when minor changes in the
$p(\psi)$ functions occur.

2.1.6   Iteration Levels

In all SCM TF calculations there are two main iteration levels :
a) mass flow iteration (outer iteration);
b) streamline iteration (inner iteration).
This is particularly true in the case of the direct problem; in the inverse
or design problem the streamline iteration can be avoided if the specific mass
flow distribution is chosen as the design law (see [2.25]).
The nesting of the iteration levels may be accomplished in many ways, and
several iteration schemes have been proposed. Some points to be noted on such
schemes are :

2.1.6.1   Mass Flow Iteration

Two different iterative procedures are possible, depending on the REE form used.
With form (F1), the mass flow iteration is performed by adjusting the velocity level on a reference streamline, while with form (F2), the pressure level
is adjusted.
As observed by Novak [2.15], problems are posed by choked blade passages
with some subsonic and some supersonic streamtubes.
Indeed, these problems exist if the boundary conditions of type (a) in
section 2.1.4 are selected. In fact, the mass flow rate no longer increases in
choked channels where the total pressure is held fixed and the (discharge) static
pressure lowered; also if the discharge pressure and temperature are fixed and
total pressure increased, a unique relation between (total) pressure and mass
flow rate still holds. On this basis the boundary conditions (b) in section 2.1.4
have been adopted, as already suggested in [2.13].

The solution for a given mass flow rate can be better obtained by form (F2) of REE, the pressure being the quantity to be corrected upstream, on the reference streamline.

The pressure corrections necessary to obtain the desired value of the mass flow rate, should be done once the REE has been integrated in all calculation stations.

Experience shows that the up-dating of the pressures is one of the most critical aspects of a TF calculation; the following method has proven reliable for most cases.

Firstly, it was supposed that the mass flow m_k at the k-th station is depending only on the pressure p_k on the reference streamline at the same station and on the pressure p_{k-1} at the (K-1)-th station upstream :

$$m_k = m_k(p_{k-1}, p_k) \tag{2.1.28}$$

Defining an operator δ as the difference NEW-OLD in an iterative process, on has :

$$\delta m_k = \frac{\partial m_k}{\partial p_{k-1}} \delta p_{k-1} + \frac{\partial m_k}{\partial p_k} \delta p_k \tag{2.1.29}$$

and then

$$p_{k-1}^{NEW} = p_{k-1}^{OLD} + \frac{\partial p_{k-1}}{\partial m_k} \left(\delta m_k - \frac{\partial m_k}{\partial p_k} \delta p_k \right) \tag{2.1.30}$$

where

$$\delta m_k = m - m_k^{OLD} \tag{2.1.31}$$

is the error in mass flow rate.

The pressure correction formula (2.1.30) works recursively starting from the last station where, as specified, the discharge pressure is fixed and then $\delta p_k = 0$.

The evaluation of partial derivatives may be done approximately by the usual monodimensional isentropic gas flow relationships :

$$\frac{\partial m_k}{\partial p_{k-1}} = \frac{m}{c_k^2 \rho_{k-1}} \tag{2.1.32}$$

$$\frac{\partial m_k}{\partial p_k} = \frac{m}{c_k^2 \rho_k} \left[M_k^2 - 1 \right] \qquad \text{for } M < 1$$

$$\tag{2.1.33}$$

$$\frac{\partial m_k}{\partial p_k} = 0 \qquad \text{for } M \geqslant 1$$

Some points should be emphasized :
a) for transonic blades, the flow deviation angle must be taken into account properly at each iteration, and the derivative $\partial m_k / \partial p_k$ must be set to zero for $M_k \geqslant 1$, otherwise the process does not converge to the true solution;
b) for transonic meridional flow in annuli, the process may not converge, because a solution cannot be obtained if a throat is not specified;
c) in order to have a stable iterative process, partial derivatives (2.1.32) and (2.1.33) may be somewhat overestimated by means of an empirical factor.

2.1.6.2 Streamline Iteration

In the direct TF problem a first guess of the streamline pattern should be made, and then updated at each iteration.

The new radial position of each streamline can be easily obtained from the calculated specific mass flow distribution, by integration of

$$r dr = \frac{1}{2\pi\rho C_z (1-\sigma\tan\gamma)} d\psi \qquad (2.1.34)$$

The use of a damping factor is common practice to overcome the instability caused by the change in streamline slope and curvature terms appearing in REE. Detailed discussions about the choice of good damping factors have been developed by Novak [2.15], Wilkinson [2.27,2.28]. A slight improvement suggested here is the use of a damping factor on the streamtube area instead of on the radial streamline position :

$$r_{NEW,DAMPED} = \sqrt{r_H^2 + d(r_{NEW}^2 - r_H^2) + (1-d)(r_{OLD}^2 - r_H^2)} \qquad (2.1.35)$$

This seems closer to the physics of the considered flow.

A substantial decrease in the required number of iterations and then of the computer time may be obtained with the use of a "dynamic damping factor". The basic idea is that during early iterations the process is much more likely to be unstable than when near convergence.

Because constant damping factors have to be small, as required by the more unstable phase of the process, this results in the slow progress towards convergence.

As a rough estimate of the potential instability, the maximum streamline error e_S at each iteration may be compared with the prescribed error e_p to be reached; a dynamic damping factor can then be defined as :

$$d = e_p/e_S, \qquad\qquad d \leqslant d_{max} \qquad (2.1.36)$$

with d_{max} given as input.

In this way the total iteration number may be decreased by more than one half.

As before, the streamline error e_S should be evaluated bearing in mind the axisymmetric character of the flow; a convenient relative streamline error may be defined as

$$e_S = \frac{r_{NEW}^2 - r_{OLD}^2}{r_T^2 - r_H^2} \qquad (2.1.37)$$

The shift of the outermost streamlines are appropriately weighted because of their larger influence on the solution.

A point seldom emphasized in the literature is the influence of the streamline distribution on the calculated mass flow rates for fixed reference pressure (or velocity) in a station. Indeed most of the methods proposed perform the mass flow iteration with fixed streamline positions. Now it can be easily shown (Fig. 2.1.3) that, for high expansion ratio blades, as those encountered in low pressure sections of large steam turbines, the radial distribution of mass flow - and hence the streamline position - is highly influenced by the expansion ratio itself (see for instance Troilo [2.29]). So the reference pressure (or velocity) updating could be made meaningless by the streamline displacement, especially in the early iterations when the streamlines undergo the largest radial shifting. The practical outcome of the above fact is the necessity to use a reduced damping

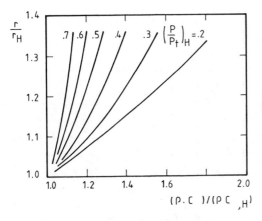

Curves have been obtained applying the isentropic simple radial equilbirum,

$$\frac{1}{\rho}\frac{\partial p}{\partial r} = \frac{c_\theta^2}{r} \text{ , to a stator blade having a}$$

flow angle described by $\tan\alpha = \frac{r}{r_H}\tan\alpha_H$.

Similar curves are obtained for different radial distribution of flow angle. The radial increase of the specific mass flow corresponds to a radial shifting of the streamlines.

FIGURE 2.1.3   Influence of expansion ratios on radial mass flow distribution

factor, so increasing the computer time.

A more advisable procedure for meridional TF, where the change in passage area is a function of the radius change squared, consists of an inner iteration on the streamlines, during the REE integration, to obtain a radial solution consistent in terms of specific mass flow distribution. In this way, of course, the slope and curvature terms are held fixed and changed only at an outer iteration level.

In other words, a given expansion ratio affects the mass flow passing capability of a blade row in two ways :
a) via the velocity level;
b) via the specific mass flow distribution
and both should be taken into account.

In conclusion, therefore, the following iteration scheme is recommended :
- guess the pressures on the reference streamline;
- guess the streamline pattern;
- integrate the REE, updating the stream function in the current station to get the radial mass flow distribution required by the pressure ratio;
- once all stations have been treated, update the reference streamline pressures to get the required mass flow rate, renew the slope and curvature terms on the basis of the new streamlines, and go back to the previous point until convergence is obtained.

2.1.7   Streamline Curvature Calculation

Considering a streamline as a function $r(z)$ defined numerically by means of a finite number of points in the (r,z) plane, all pertaining to the given streamline, the first and second derivative of such a function have to be evaluated, whichever REE form is used.

This problem was given careful attention and deep analysis, among others, by Shaalan and Daneshayer [2.30] , Wilkinson [2.28] , Renaudin & Somm [2.11] : the cited papers give detailed information about various calculation methods and their precision. Most computer codes today use the double-spline method, which has been found very suitable since its presentation [2.30] .

However, it is worth mentioning briefly the temporal-spline method proposed in [2.11] , which is unique and has been compared with a simple second order difference formula in a work by Nurzia & Troilo [2.31] where non negligible influence of the different methods on the overall solution was shown (Fig. 2.1.4).

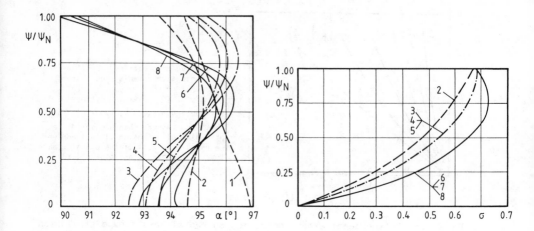

1 : simple radial equilibrium with no curvature
2 : finite difference formulas
3,4,5 : temporal streamline curvatures [2.11] with no blockage effects, and different weights
6,7,8 : the same as before with a blockage effect accounted for

FIGURE 2.1.4   Absolute outlet flow angle α and streamline slope σ versus stream
    function ψ, at inlet of a low pressure steam turbine stage (from [2.31])

To avoid convergence problems that tend to arise when dealing with steam turbine
last stages, the authors applied a progressive weighting of the first and second
derivative terms in the REE by a factor from zero to one. Extensive calculations
performed with the double-spline method and variable derivative weighting have
shown marked changes in the solutions for weighting values around 0.6, and minor
effects were found for higher values.
    This observation demonstrates the difficulties in attributing "true" deriva-
tives to the streamlines. In spite of the great success of SCM, this problem
remains unsolved.
    The stability and accuracy properties of numerical methods for finding the
second derivative of a curve have been treated in a general mathematical sense
in the papers [2.15] and [2.27].
    In the application of SCM to TF calculation of multistage machines, there is
a particular source of instability that is worth mentioning.
    A good design rule is to have nearly constant static pressure at the rotor
blade outlet, because this condition leads to a uniform load distribution along
the stage, which is generally desirable. Where high slopes and curvature exist,
as in the low pressure turbine final stages, and especially in the calculation
stations located at the turbine outlet, a constant pressure corresponds to a
vanishing difference of non-zero terms in the right hand side of equation (2.1.9)
or equivalent equations. As a consequence, small errors or uncertainties in the
curvature term computation result in a large error on a term which would be near
to zero. This fact causes a particular form of instability that has not been
adequately reported previously. This instability may be avoided by the appropriate choice of calculation stations; so the rule of orthogonality, which elsewhere
allows a reduction of the weighting of some terms, may not be the best and selection of options where a non-zero pressure gradient exists, may be preferable.

2.1.8 Loss and Flow Angle Calculation

While the choice of a proper technique for the mass flow iteration and the stream-line curvature estimate is decisive for a successful numerical solution, the correct estimate of losses and flow angle deviation is the key to obtain a reli-able solution as far as the machine performance is concerned.

In this sense a direct TF calculation is a powerful design tool, in that it permits an accurate prediction of the machine performances, and then an evalua-tion of different design criteria with respect to their impact on the efficiency and characteristic curve.

2.1.8.1 *Loss Calculation - Profile and Secondary Losses*

The best results are obtained by calculating these losses as functions of the flow parameters, for a given blade geometry, and updating them at each iteration. For this procedure on-line loss correlations should be available. These can be obtained from general loss correlations as, for example, that of Craig & Cox [2.32], provided that preliminary calculations are done for selected blade sec-tions at given heights. Alternatively losses can be obtained from experimental data available to the computer code user for his particular blade shapes. Since the blade sections are defined at fixed radii, while the streamline position varies at each outer iteration, interpolation procedures are required (a) to determine the flow conditions at the radii at which the blade geometry is known, in order to perform the loss calculations and (b) to recalculate the losses at the radii of the streamlines.

Among the available general loss correlations, that of Craig & Cox [2.32] is widely used even though a slight overestimate has been found in off-design conditions (see e.g. Rigoli & Troilo [2.33], Macchi [2.34]).

Loss correlations are commonly referred to plane cascades, whilst in large turbines an important sweep effect exists because of large radial velocity components.

As suggested by Smith [2.35] and discussed by e.g. Starken [2.36] and Dzung [2.37], flow properties affecting losses can be expressed as axial and tangential velocity components, as long as blades are not tilted.

From the computational point of view, loss correlations can be given either in analytical form, or in the form of tabulated values. The second method is preferable because it has a least two advantages :
a) higher flexibility as regards the loss data sources;
b) loss correlation management is separate from that of equation encoding.

The Craig & Cox correlation [2.32] can be used in the form of tabulated values as follows :
- profile losses are expressed as a fraction ξ_p of the available isentropic head

$$\xi_p = \xi_{p,b} \cdot c_{Re} \cdot c_i \cdot c_{te} + \Delta\xi_{p,te} + \Delta\xi_{p,e} + \Delta\xi_{p,M} \qquad (2.1.38)$$

where

$\xi_{p,b}$ is called "basic profile loss"

c_{Re} is a correction factor depending on Reynolds number and surface finish

c_i is a correction factor due to incidence angle

c_{te} is a correction factor due to trailing edge thickness

$\Delta\xi_{p,te}$ loss increment due to trailing edge thickness

$\Delta\xi_{p,e}$ loss increment due to back radius

$\Delta\xi_{p,M}$ loss increment due to Mach number.

A careful examination of the Craig & Cox correlation shows that, for given blade sections, arranged in given rows (i.e., with given pitch/chord ratio), for given surface finish :

$\xi_p \cdot c_i$ is function only of the incidence flow angle

c_{Re} is function only of the Re number

c_{te} is function only of the outlet flow angle

$\Delta\xi_{p,te}$ is a constant

$\Delta\xi_{p,e} + \Delta\xi_{p,M}$ is a function only of the Mach number.

In this way, for each blade section a few (four) one-entry tables and one constant may be calculated once and for all and stored, so that loss calculation reduces to simple interpolation.

- secondary losses are similarly expressed as

$$\xi_s = \xi_{s,b} \cdot c_{Re} \cdot c_A \qquad\qquad\qquad (2.1.39)$$

where

$\xi_{s,b}$ is a "basic secondary loss"

c_A is a correction factor depending on the blade aspect ratio.

As above, for a given blade section :

$\xi_{s,b}$ is function only of the inlet/outlet kinetic energy ratio

c_A is function only of the aspect ratio

and the same situation as for profile losses holds.

2.1.8.2 Loss Calculation - Annulus Losses

These losses are expressed in practice as a fraction of the leaving kinetic energy from manufacturer's data.

2.1.8.3 Miscellaneous Losses

Under this name are commonly included losses of very different kind, such as disk windage, leakage and lacing-wire losses.

Of course, these losses are not relevant in the blade design, which is more concerned with profile and secondary losses and, to some extent, with the tip leakage. Nevertheless, the accurate calculation of these miscellaneous losses is very important from the manufacturer's and customer's point of view, because it defines, together with mechanical losses, the net power output at the turbine shaft coupling. Furthermore, the amount of these losses is comparable with that of the profile and secondary ones, and they can often be reduced significantly by careful mechanical design.

In this respect, there are well-known trade-offs, for example, between the mechanical reliability obtained with a lacing wire and the power losses it causes, as well as between the increased shaft length due to larger axial spacings and the associated reduction in leakage and windage losses.

From the whole calculation point of view, it should be mentioned that leakage and windage losses are connected to flows that come into and out of the meridional flow plane, so that they interact with the main meridional flow in a number of ways (Fig. 2.1.5). No attempt has been made to obtain a fully consistent solution for the branched flows; it seems advisable to introduce the leakage flow once a first solution has been obtained, and then to obtain an improved solution with the correspondingly modified flows through the blades.

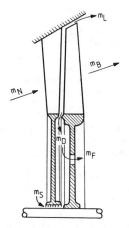

m_N main flow through the stator nozzles
m_B main flow through the rotor blades
m_L tip leakage flow
m_D leakage flow to the wheel disk
m_F balance hole flow
m_S shaft leakage flow

FIGURE 2.1.5 Meridional section of a turbine stage with main and leakage flows

2.1.8.4 *Flow Angle Calculation*

A correct flow angle estimate is important, in TF calculations, as much as the correct loss evaluation, for at least two reasons :
a) flow angles in large steam turbines are limited to some 15-20 degrees for the stator nozzle and some 20-30 degrees for running blades, in order to have highly loaded stages, so reducing their numbers for a given overall enthalpy drop. Small absolute errors in flow angle evaluation therefore correspond to significant relative errors, which directly affect the estimate of the mass flow rate passing capability of the turbine.
The consequence of errors in flow angles is a shift in the predicted characteristic curve, which can introduce serious problems : either the desired mass

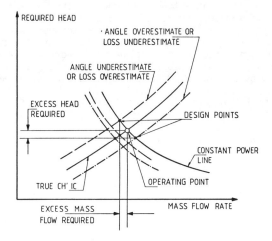

Figure 2.1.6 An error in loss and/or flow angle prediction causes always a power reduction, unless the head or mass flow rate is increased beyond the design point

cannot pass with a given pressure drop, or it passes with a reduced head, so that in both cases the turbine cannot give the expected power unless the matching steam generator is forced in operation beyond its nominal design point (Fig. 2.1.6).
b) the error in the required head evaluation arising from the error in flow angles affects in turn the loss estimate, especially in the supersonic range where losses depend strongly on Mach number. As well as giving incorrect evaluation of the turbine efficiency due to wrong loss prediction, it should be pointed out that the angle errors result also in a difference in losses which shift the predicted characteristic curve, with the same consequence as discussed under (a). Both effects may be cumulative.

Reliable flow angle correlations should take into account the cascade geometry, the annular flow effects and efficiency and Mach number values; a careful description of the transition region around M = 1, where the largest deviations between geometric and flow angles take place, should also be considered (see e.g. [2.12,2.38,2.39]).

From the calculation point of view, as with the losses, flow angles are better evaluated between two outer iterations, when all necessary data are comtemporarily available.

While an updating of the losses may be done once every few iterations in order to save computer time, flow angles should be updated at each iteration because the supersonic deviation commonly encountered in large steam turbines directly affects the convergence of the mass flow iteration process.

2.1.9 Wetness Effects

Thermal cycles for power generation purposes, both in fossil - and nuclear - fueled steam power plants, incorporate an expansion line penetrating the saturated region. The appearance of the liquid phase is connected to several phenomena, among which the two most important may be summarized as :
a) the formation of droplets, whose paths are different from the main gas flow streamlines, in such a way that they hit the blade surfaces causing erosion;
b) the formation of undercooled steam in a metastable state, which condenses suddenly, producing an important thermodynamic loss.
Many other aspects could be listed, such as liquid film deposit on blades, water centrifugation, fog entrainment; in the following the condensation shock will be considered because of its substantial influence on the main flow. Further theoretical treatments of wetness are given in §5.1.

The condensation shock is indeed a complex phenomenon which starts from condensation nuclei around which droplets are growing towards an equilibrium state; it has been described in great detail in the basic work of Gyarmathy [2.40].

The real process takes place in finite time, but it has been considered as a local discontinuity by Gyarmathy [2.41] which is a convenient approximation for present purposes.

The equations of the condensation shock located on a surface orthogonal to meridional streamlines are sufficiently simple to be included in a TF calculation code, where they account for :
- conservation of tangential momentum :

$$C_{\theta_{bs}} = C_{\theta_{as}} \tag{2.1.40}$$

(here bs and as mean respectively "before shock" and "after shock").
- continuity :

$$\rho_{bs} C_{m_{bs}} = \rho_{as} C_{m_{as}} \tag{2.1.41}$$

The density ρ_{bs} should be calculated for a superheated steam expansion, while ρ_{as} is the equilibrium density.

- energy conservation :

$$h_{bs} + \frac{c_{bs}^2}{2} = h_{as} + \frac{c_{as}^2}{2} \tag{2.1.42}$$

Here also enthalpies h_{bs} and h_{as} should be calculated for superheated and equilibrium state respectively.
- conservation of momentum normal to streamline :

$$p_{bs} + \rho_{bs} c_{m_{bs}}^2 = p_{as} + \rho_{as} c_{m_{as}}^2 \tag{2.1.43}$$

By appropriate arrangement of these equations, and a suitable iterative numerical procedure, it is possible to calculate the conditions before the shock for given conditions after the shock. When all these are known, the wetness loss may be calculated as :

$$\xi_W = 1 - \frac{h_t - h_{as}}{h_t - h_{bs}} \tag{2.1.44}$$

where h_t is the total enthalpy upstream, and the deviated flow angle is

$$\alpha = \text{atan} \frac{c_{z_{as}}}{c_{\theta_{as}}} \tag{2.1.45}$$

with $c_{z_{as}} = c_{m_{as}} / (1+\sigma^2)$.

In this way, during the REE integration, the dependence of the function f (see Eqn. 2.1.23) on p, also takes into account the condensation process. However, this method permits the treatment of the condensation shock on the assumption that it is located at the blade trailing edges.

Calculation experience shows that, especially for high expansion ratios, conditions are often encountered at the blade outlet for which the mathematical solutions of shock equations does not exist. In such cases, the shock probably takes place inside the blade channel but its position cannot be found without detailed calculations of the actual condensation process. This is an important limitation as regards the treatment of the shock in a concentrated form; in particular, the disappearance of the shock causes a large discontinuity in the flow angle predicted by equation (2.1.45), giving large deviations which are near the limit of possible solution. More detailed calculation methods, compatible with TF main equations are required.

In spite of the above mentioned limitations, improved agreement with experimental data has been found in TF calculations of wet steam turbines using the concentrated condensation shock equations, and positioning the shock at the blade outlet whenever possible.

If the shock cannot be positioned at the blade outlet, i.e., if the mathematical shock equations have no solutions, a wetness loss has been taken into account by an empirical formula

$$\xi_W = 1 - \left(\frac{p}{p_t}\right)^a \tag{2.1.46}$$

where $a \approx .054$ and p, p_t are respectively downstream static and upstream total

pressure. This relation is in good agreement with the losses calculated by means
of the shock equations.

2.1.10 Application to a Last Stage Design

As described in chapter 1, the last stage of a large steam turbine is difficult
to design for a number of reasons, such as :
a) it has the larger tip to hub radius ratio and hence the strongest three-
dimensional effects;
b) it operates in wet steam;
c) flows are as a rule transonic and supersonic.
    The analysis of the problems that arise, and the calculation of particular
aspects of the flow should be performed by appropriate methods.
    The TF meridional calculation provides an insight on the stage behaviour.
It provides, for example :
a) the pressure ratio of the stage;
b) the radial Mach number distributions;
c) the radial pressure distribution;
d) the incidence of the various sources of losses;
e) the flow angles at the blade outlets
enabling the following design features to be evaluated :
a) the compatibility with the upstream stages;
b) the efficiency figures obtained;
c) the mass flow passing capability;
d) the operation at off-design.
    A number of inputs are also obtained for particular calculations :
a) the velocity distribution at inlet and outlet of the blades for secondary
flow calculation;
b) the pressure drops at blade ends for leakage flow calculation;
c) the Mach numbers, for the choice of blade-to-blade calculation methods for
the design refinement of the blade shapes already defined.
    Sample calculations were done for the ANSALDO 320MW, LP turbine described in
§6.2. Full geometrical details are given in [2.22].   Last stage calculation
results are presented in figure 2.1.7. Results are presented with and without
the inclusion of the condensation shock calculations showing the difference
obtained. All quantities are plotted versus H*, the non-dimensional blade height,
defined as :

$$H^{\star} = \frac{r - r_H}{r_T - r_H}$$ (2.1.47)

    The actual height of the stage is about 840 mm. Solid lines refer to shock
inclusive calculations, following the method outlined in section 2.1.9; dotted
lines refer to calculations with the steam in equilibrium throughout. Blade
angles, for comparison with flow angles, are indicated by full lines marked with
small triangles. The subscript 1 is used in figure 2.1.7 to indicate the fixed
blade outlet station, while the subscript 2 indicates the stage outlet station.
    Figure 2.1.7-a shows the axial Mach number at the stage outlet (M_{2ax}), which
is well below unity, the inherent limit of SCM; the parabolic-type behaviour of
M_{2ax} is a design choice and, as can be seen, is little affected by the steam
equilibrium assumptions.
    Figure 2.1.7-b shows the static pressure at the stage outlet (p_2), and it can
be seen how the non-equilibrium calculation gives a more uniform distribution,
closer to the desired constant value; thus, in this respect, stage design corrections based on the equilibrium analysis would be inappropriate. At nearly
half blade height the static discharge pressure has the same value both for equilibrium and non-equilibrium case : this is merely because the static pressure is

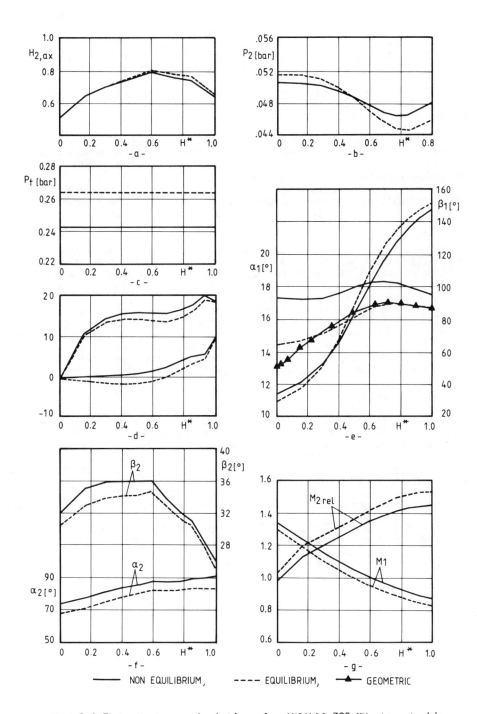

FIGURE 2.1.7 Last stage calculations for ANSALDO 320 MW steam turbine

imposed at that point as a boundary condition (see § 2.1.4).

Figure 2.1.7-c shows the stage inlet total pressure (p_0). Since the mean level of outlet pressure is the same for both calculations, figure 2.1.7-c shows one of the most important results, i.e., the non equilibrium calculation gives a lower stage pressure ratio. That is, the superheated expansion and occurrence of condensation shock results a higher mass flow passing capability. Alternatively the stage requires a lower inlet pressure level to pass a given mass flow. This result, coupled with the flow angle variations seen in the next figures, is relevant because of its consequence for the calculation of power output of the turbine plant.

Figure 2.1.7-d shows the streamline slope at the outlet of the fixed (σ_1) and running (σ_2) blades. The equilibrium and non-equilibrium results give the differences in the streamline pattern due to the different calculated operating conditions, especially the pressure ratio.

Figure 2.1.7-e shows the absolute (α_1) and relative (β_1) flow angles at the fixed blade outlet, compared with the geometric value given, as is customary, as arsin (throat/pitch). Calculation in equilibrium condition gives only the supersonic flow angle deviation, which is a maximum at the hub where the highest Mach number occurs, followed by a small overturning in the region of unity Mach number (cfr. Fig. 2.1.7-g). Calculation at the non-equilibrium state gives also a large flow deviation associated with the strong condensation shock located at the blade outlet. A similar effect may be seen in figure 2.1.7-f for the relative (β_2) and absolute (α_2) flow angle at the stage outlet (geometric angles were not available for presentation).

Finally, figure 2.1.7-g shows the absolute (M_1) and relative (M_{2rel}) Mach numbers, respectively, at the fixed blade and running blade outlet. Whilst the outlet Mach number evaluated in non-equilibrium is lower, as could be expected because of the lower pressure ratio, the absolute Mach number at the fixed blade outlet is higher when evaluated in non-equilibrium. This indicates a significant difference in the calculated stage reaction degree in the two cases.

2.1.11 Conclusions

The meridional through-flow calculation for low pressure steam turbines, using the streamline curvature method may be performed with different forms of the same equation. To obtain a reliable design tool it is also necessary to select :
1) appropriate boundary conditions;
2) a suitable numerical integration scheme, observing that in some cases the radial equilibrium equation fails to match the Lipschitz condition for a stable integration process;
3) the best arrangement of iteration levels and of the algorithms for up-dating the guessed values;
4) a suitable scheme of streamline slope and curvature calculation;
5) reliable loss and flow angle correlations;
6) a model for the treatment of the wetness effects.

Detailed account of SCM theory can be found in the cited references. Some critical aspects related to the above problems have been presented and discussed, giving hints for refinement and improvement of available codes.

Finally, a sample case of a last stage design has been presented, and the importance of taking into account wetness effects has been emphasized.

2.2 MATRIX THROUGH FLOW SOLUTION

D.H. Evans

2.2.1 Basic Equations

The purpose of a flow field code is top optimize the thermal performance of the steam turbine by accurately defining the flow field properties and minimizing losses by giving the designer considerable flexibility when developing his design. Failure to define the flow field accurately results in "mismatching" of stationary and rotating blades leading to flow incidence as shown on figure 2.2.1. If the flow enters the blade at a direction different from the minimum loss direction, increased losses result.

If the deviation is sufficient that the increased losses are measurable, the performance penalty may be economically significant to a turbine operator. To illustrate the importance of defining the flow field properties accurately, the increased losses occurring due to flow incidence are shown on figure 2.2.2. Note the loss increase in the tip region of a last row rotating blade of a low pressure steam turbine for a rather modest change in flow inlet angle.

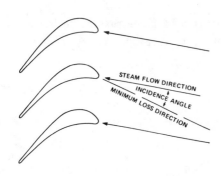

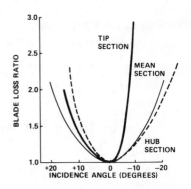

FIGURE 2.2.1 Definition of flow
incidence angle

FIGURE 2.2.2 Blade section loss
as a function of incidence

A number of complex computer codes have been developed by various investigators to solve the meridional flow field. Some codes have been extensively applied to the design and analysis of turbomachines. Verification of the codes requires testing at considerable expense to the manufacturers. The two most common techniques for determining the flow field are the streamline curvature method and the matrix through flow method. Both methods assume axisymmetric conditions throughout although sometimes a blockage term is introduced to account for blade thickness.

The streamline curvature method has been described in section 2.1; this section considers the matrix through flow solution.

The matrix approach utilizes a stream function satisfying the equation of continuity. A matrix of grid points is placed throughout the blade path and the stream function determined by a numerical process.

The momentum equations in the meridional plane are :

$$C_r \frac{\partial C_r}{\partial r} + C_z \frac{\partial C_z}{\partial z} = -\frac{1}{\rho}\frac{\partial p}{\partial z} \qquad (2.2.1)$$

$$C_r \frac{\partial C_r}{\partial r} + C_z \frac{\partial C_z}{\partial z} - \frac{C_\theta^2}{r} = - \frac{1}{\rho} \frac{\partial p}{\partial r} \tag{2.2.2}$$

Only one of the two equations must be solved and customarily equation (2.2.2) is chosen. This equation is referred to as the radial equilibrium equation. The definition of rotation in cylindrical coordinates is :

$$\frac{\partial C_z}{\partial r} - \frac{\partial C_r}{\partial z} = - 2\omega_\theta \tag{2.2.3}$$

and defining a stream function from the continuity equation gives :

$$C_z = \frac{1}{br\rho} \frac{\partial \psi}{\partial r} \tag{2.2.4a}$$

$$C_r = - \frac{1}{br\rho} \frac{\partial \psi}{\partial z} \tag{2.2.4b}$$

where   b is a blockage term, equal to the ratio of blade passage to blade pitch. Substituting equations (2.2.4a) and (2.2.4b) into (2.2.3) results in :

$$\frac{\partial \left(\frac{1}{br\rho} \frac{\partial \psi}{\partial r} \right)}{\partial r} + \frac{\partial \left(\frac{1}{br\rho} \frac{\partial \psi}{\partial z} \right)}{\partial z} = - 2\omega_\theta \tag{2.2.5}$$

   From the first law of thermodynamics :

$$\frac{1}{\rho} \frac{\partial p}{\partial r} = \frac{\partial h}{\partial r} - T \frac{\partial s}{\partial r} \tag{2.2.6}$$

and differentiating for the energy equation :

$$h_t = h + \frac{C_\theta^2 + C_z^2 + C_r^2}{2}$$

$$\frac{\partial h_t}{\partial r} = \frac{\partial h}{\partial r} + C_\theta \frac{\partial C_\theta}{\partial r} + C_z \frac{\partial C_z}{\partial r} + C_r \frac{\partial C_r}{\partial r} \tag{2.2.7}$$

Substituting (2.2.6) and (2.2.7) into (2.2.2) :

$$\frac{\partial h_t}{\partial r} - C_\theta \frac{\partial C_\theta}{\partial r} - C_z \frac{\partial C_z}{\partial r} - C_r \frac{\partial C_r}{\partial r} - T \frac{\partial s}{\partial r} = \frac{C_\theta^2}{r} - C_z \frac{\partial C_r}{\partial z} - C_r \frac{\partial C_r}{\partial r}$$

$$\frac{\partial h_t}{\partial r} - T \frac{\partial s}{\partial r} - C_\theta \frac{\partial C_\theta}{\partial r} = \frac{C_\theta^2}{r} + C_z \frac{\partial C_z}{\partial r} - C_z \frac{\partial C_r}{\partial z}$$

$$\frac{\partial C_z}{\partial r} - \frac{\partial C_r}{\partial z} = \frac{1}{C_z} \left(\frac{\partial h_t}{\partial r} - C_\theta \frac{\partial C_\theta}{\partial r} - T \frac{\partial s}{\partial r} - \frac{C_\theta^2}{r} \right)$$

which leads to the following :

$$\frac{\partial}{\partial r}\left(\frac{1}{br\rho}\frac{\partial\Psi}{\partial r}\right) + \frac{\partial}{\partial z}\left(\frac{1}{br\rho}\frac{\partial\Psi}{\partial z}\right) = \frac{1}{C_z}\left(\frac{\partial h_t}{\partial r} - \frac{C_\theta}{r}\frac{\partial(rC_\theta)}{\partial r} - T\frac{\partial s}{\partial r}\right) \qquad (2.2.8)$$

This is the basic flow field equation which can be solved by numerical techniques via utilization of a large scale digital computer. The above equation is written in finite difference form and the stream function is determined by a relaxation method. This requires that a large number of grid points be superimposed on the turbine flow path and the field equation solved iteratively until convergence is achieved.

The numerical technique is a relatively simple five point star resulting in a finite difference equation of the form :

$$\Psi_{i,j} = A_1\Psi_{i+1,j} + A_2\Psi_{i,j+1} + A_3\Psi_{i-1,j} + A_4\Psi_{i,j-1} + A_5$$

The coefficients are a function of the grid spacing and the thermodynamic properties at each of the grid points.

Equation 2.2.8 is structured for what is normally referred to as the "design" problem. The right hand side of the generalized flow field equation gives some insight into the basic turbine design approach. Following the selection of turbine size and number of stages, determined by power output and nominal efficiency requirements, the design begins with the flow field calculations. With an initial assumption of turbine boundaries, the work distribution is varied along the height of the blade. The work variation is defined by the first two terms (h_t and C_θ) while the third represents loss distributions embodied in the familiar entropy function. The loss distributions, based on experimental data ranging from linear cascade tests to full size turbine tests, become an integral part of the design solution. Optimum performance is approached by balancing work distribution (blade loading) against the resulting blade losses. During the initial design process the final blade shapes are unknown as well as the absolute losses. Therefore, generalized loss functions based on flow turning and Mach number are used during this design phase. Although absolute levels of loss are not known at this stage the generalized loss functions are adequate to approach optimum performance.

The streamline curvature is included in the left hand side of the equation. As work distributions are varied and balanced against losses, there results a variation in mass flow distribution and correspondingly streamlines, especially

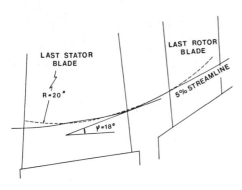

FIGURE 2.2.3 Streamline curvature
in last stage

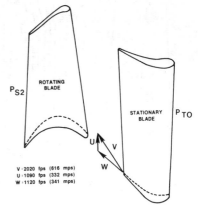

FIGURE 2.2.4 Velocity triangle
at stator base

near the turbine boundaries. Achieving high thermal performance near the base of
the last stage of a low pressure turbine is extremely difficult. The flow leaves
the last stator at high Mach number and then passes through the rotor blade which
must turn the flow 80 to 100 degrees. Figure 2.2.3 shows typical velocity triangles for these conditions. Reducing the stator exit Mach number by raising the
pressure can improve the blading performance. Some benefit can be achieved by
introducing streamline curvature in the base region as shown in figure 2.2.4.
The curvature can be achieved by proper selection of the boundary shape and the
mass flow distribution.

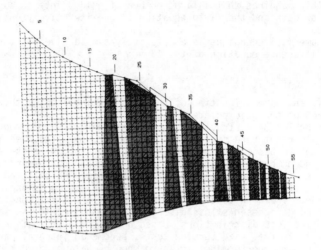

FIGURE 2.2.5 Blade path grid system

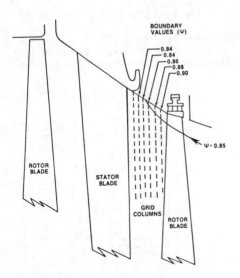

FIGURE 2.2.6 Extraction flow
              modeling

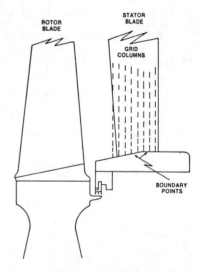

FIGURE 2.2.7 Stator blade wall
              boundary

2.2.2   The Design Method

Boundary values must be specified everywhere to solve the design flow field
equation. Figure 2.2.5 shows a five stage fossil turbine including a diffusing
type passage at the turbine exhaust. Grid points are placed throughout the turbine and fall inside the blade passages and in the spaces between. Stream
functions are specified at the inlet and exit boundaries which are spaced sufficiently from the blading to not influence results at the blade leading and
trailing edges. Normally the exit boundary is placed downstream approximately
half the length of the last rotor blade. Usually constant mass flow distribution
is assumed at these boundaries but the designer has the option of other distributions. This is especially useful when "patching" two parts of a turbine together
in separate calculations. The outer boundary values are defined by the flow
extraction arrangement. Figure 2.2.6 shows the region between a rotor blade and
stator blade where flow is extracted for feedwater heating. The stream function
value changes from rotor exit to stator inlet in the amount equal to the flow
removed.

The boundary shape is defined primarily by the inner and outer radii of the
blades. Additional points are added to define the boundary more accurately.
Figure 2.2.7 shows a last stage where additional points are included inside the
stationary blade.

Once the blade path is defined and the blade leading and trailing edges
located the calculation procedure is started. The grid system is automatically
generated after specifying the number of columns and rows. Initial values of
tangential velocity distribution (C_θ) are necessary along with blade blockage and
loss data. Much of this data is filed in the computer so that subsequent runs
require a minimum of data management. During the relaxation process of solving
the flow field equation the entropy function is continually updated. The density
is revised during the numerical process via the ASME steam table computer codes
[2.42]. Convergence is achieved when changes in stream function are sufficiently
small to have negligible effects on the flow field properties. Once convergence
is achieved the designer may select another distribution of tangential velocities
to optimize his design.

Special consideration must be given to blade losses near the "endwalls"
where the blades are attached to the cylinder and rotor. Secondary flows due to
the interaction of wall and blade boundary layers result in increased blade losses.
These losses are significant in the hub regions, especially in the last stage of
a large fossil low pressure turbine. For the flow field design solution, coefficients are introduced to increase the normal blade loss. The coefficients are
multipliers applied to the blade loss data. The coefficient values are empirical
based on numerous turbine tests where flow field properties have been determined
by internal measurements with traverse probes. Extensive test data from a low
pressure research turbine proved invaluable during the development of the matrix
method computer code. The "endwall loss" coefficients developed at the time have
been only slightly modified as additional test data were accumulated from several
different designs including fossil and nuclear turbines.

2.2.3   Verification of the Analytical Method

The matrix through flow method was first applied to a retrofit design of a large
fossil low pressure turbine. The last row blade is 31 in. (0.787 m) long with a
tip velocity of 2010 ft/s (613 m/s). Based on field test results and internal
traverse measurements on the original design, significant performance improvements were predicted for the new design. Final verification was achieved after
retrofitting several turbines with advanced design blades. The turbine flow path
is shown on figure 2.2.8.

Data was obtained from field units and laboratory facilities. Field test
results were obtained by procedures defined by the ASME Performance Test Codes.
Following the measurement of the original design thermal performance, the last

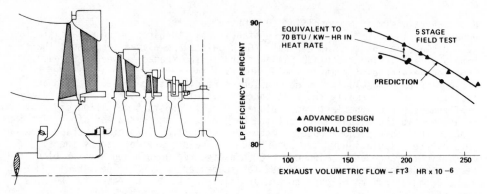

FIGURE 2.2.8   Advanced design low
                    pressure turbine

FIGURE 2.2.9 Field test results from
                original and advanced designs

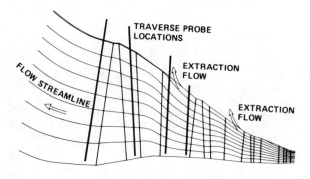

FIGURE 2.2.10   Flow streamline and traverse probe locations in test turbine

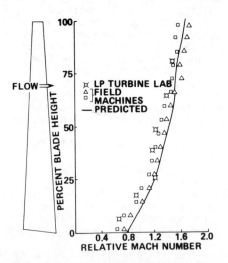

FIGURE 2.2.11   Mach number relative to last rotating blade of advanced design

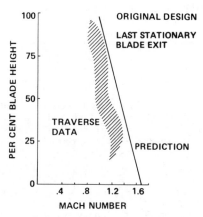

FIGURE 2.2.12 Mach number at exit of last stationary blade of original design

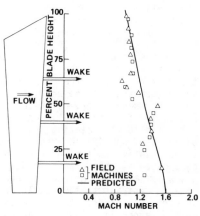

FIGURE 2.2.13 Mach number at exit of last stationary blade of advanced design

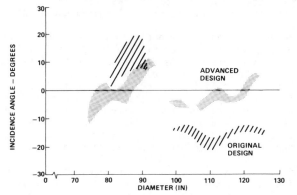

FIGURE 2.2.14 Flow incidence at inlet to last rotating blade

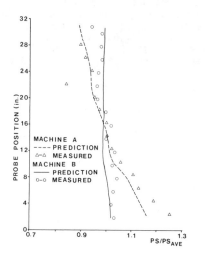

FIGURE 2.2.15 Exit static pressure - advanced design

stage blades were replaced with advanced design blades and the tests were repeated.
Figure 2.2.9 shows the improvement in low pressure turbine efficiency that is
equivalent to 70 BTU/kwh (18. Kcal/kwh) in turbine heat rate.

Internal measurements of the flow field properties by insertion of traversing
probes are essential to verify the analytical prediction method. Figure 2.2.10
shows the typical probe locations in the last two stages. Figure 2.2.11 presents
the measured Mach number leaving the last rotor blade as a function of blade
height compared to the predicted value.

The Mach number, relative to the rotating blade, is deduced from measured
properties from the stationary probe downstream of the rotor blade and then
converted to the rotating frame. Traverse data at the exit of the last stationary
blade is of special interest since the traverse results indicate the flow conditions relative to the inlet of the last rotating blade. Special care must be
excercised in this traverse location to minimize the effect of probe blockage on
the steam flow which leads to erroneous probe readings. An oblique traverse is
made in contrast to the nearly perpendicular to the machine axis traverse made
in other locations. The probe is inserted at an angle approximately 15 degrees
in the tangential direction and so sweeps across three stationary blade passages
on its way from tip to hub. The Mach numbers for both original and advanced
design configurations are shown on figures 2.2.12 and 2.2.13. The effect of
stator blade wakes can be seen clearly in figure 2.2.13. This data leads to
significant conclusions concerning the performance of the last rotating blade.
The flow incidence to the last rotating rows of the original and advanced design
turbines are seen on figure 2.2.14.

The matrix through flow design equation can be applied to a turbine of known
geometry to analyze performance at conditions different from the design point.
Such off-design conditions range from expansion line variations due to different
cycle applications to far-off-design conditions such as low load and high condenser pressure operation where the last rotating blade of a low pressure steam
turbine may be prone to stall flutter. Application of the design equation to an
analysis problem requires, in addition to convergence on stream function, convergence on a parameter unique to the turbine geometry. Typically this parameter
is flow exit angle from each blade. The flow exit angle is a function of blade
geometry and exit Mach number and this relationship is developed from cascade
tests.

Verification of the flow field calculation method has been extended to conditions different from the design point. Figure 2.2.15 presents static pressure
distribution at the last rotating blade exit. For machine A the static pressure
is highest at the base while for machine B the pressure is nearly constant. The
difference is due to operating conditions where machine A operates near design
volumetric flow and machine B much lower. Therefore, the leaving tangential velocity, or "swirl" velocity, is much higher for machine B and the static pressure
increases toward the tip due to centrifugal force. The higher static pressure
near the base for machine A is due to the effect of the diffusing flow guide which
turns the flow outward in the radial direction.

Turbine operators, when faced with an environment of escalating fuel costs,
can benefit significantly by a manufacturer refining his designs to accommodate
different condition lines. For example, figures 2.2.16 and 2.2.17 show flow incidence angle to the last stationary blade for a large 3600 RPM turbine operating
at different condition lines, noted as high and low mass flow rate per unit area
at the turbine exhaust. The basic reason for the flow to the stator to change is
a reduced rate of extraction flow for the low flow case. What appears to be data
scatter, especially for the high flow case, is also due to varying extraction
flow due to changing condenser pressure. The net effect of incidence is to increase blade losses above the minimum loss condition. By developing a family of
stationary blades for different operating conditions, the thermal performance is
improved for machines applied to steam cycles significantly different from the
design point.

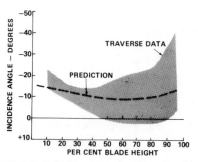

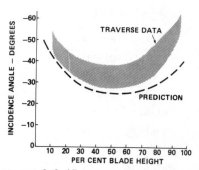

FIGURE 2.2.16 Last stationary blade incidence angle, advanced design - high flow application

FIGURE 2.2.17 Last stationary blade incidence angle, advanced design - low flow application

The analysis of turbine flows at far-off design operation is difficult, especially when flow in the hub region is reduced to a level close to where reversals are about to occur. These conditions are more important to blade reliability than performance. At very low flows negative incidence occurs at the outer portion of the last rotating blade and the blade may be susceptible to stall flutter. Criteria have been developed to design modernized blades that are much less sensitive to stall flutter than earlier designs. The ability to more precisely predict flow to the rotating blade for these conditions enhances the understanding of the flutter phenomena. Figure 2.2.18 presents predicted streamlines for the design condition and a low flow condition where we note the reduction of mass flow at the base. Figure 2.2.19 shows the change in inlet flow angle to the rotating blade.

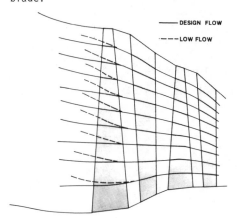

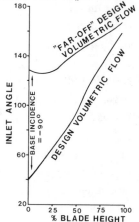

FIGURE 2.2.18 Predicted streamlines for design flow and low flow

FIGURE 2.2.19 Last rotor blade inlet flow angle at design and "far-off" design conditions

The development of a complex computer code for turbine flow field analysis requires more than an expertise in numerical analysis and general aerodynamics. Extensive experience in turbomachinery aerodynamics is essential along with a large test data base accumulated over a number of years of experimental work on models, full size laboratory units and field units. Correlation of various types of test data along with the development of a reliable loss model is necessary to calculate field properties with sufficient precision to assure high thermal performance.

NOMENCLATURE

b	blockage term (ratio of blade passage to blade width)
C	absolute velocity
c_p	constant pressure specific heat
d	damping factor
e	relative error
f	function
H	blade height
h	enthalpy
m	mass flow rate; streamline coordinate
M	Mach number
n	direction normal to quasi-orthogonals
p	pressure (static)
q	quasi orthogonal coordinate
r	radius; radial coordinate
s	entropy
t	time
T	absolute temperature
W	relative velocity
z	axial coordinate
α	absolute flow angle
β	relative flow angle
γ	angle between radial and quasi-orthogonal directions
ε	angle between streamline and radial directions
ξ	blade losses; $\xi = (C_{is}^2 - C^2)/C^2$
ρ	density
σ	streamline slope $= \text{tg}\varphi$
φ	angle between axial and streamline directions
ψ	stream function
Ω	rotational speed
ω	rotation of a fluid

Subscripts

a	after
A	aspect ratio
b	basic; before
c	curvature
H	hub
i	incidence

is   isentropic

j   referred to the j-th streamline

k   referred to the k-th station

m   meridional component

max   maximum

M   Mach

p   profile

P   prescribed

r   radial component

R   relative to rotor

Re   Reynolds

s   secondary; shock

S   streamline

t   total

te   trailing edge

T   tip

W   wetness

z   axial component

θ tangential component

REFERENCES

2.1   Wu, C.H.: A General Through-Flow Theory of Fluid Flow with Subsonic or Supersonic Velocity in Turbomachinery Having Arbitrary Hub and Casing Shapes. *NACA TN* 2388, March 1951.

2.2   Wu, C.H.: A General Theory of Three-Dimensional Flow in Subsonic and Supersonic Turbomachines of Axial-Radial- and Mixed Flow Types. *NACA TN* 2604, January 1952.

2.3   Lorenz, H. Theorie und Berechnung der Vollturbinen und Kreiselpumpen. *Zeitschrift des Vereines deutscher Ingenieure*, vol. 49, Okt. 1905, pp 1670-1675.

2.4   Smith, L.H.: The Radial Equilibrium Equation of Turbomachinery. *ASME Trans., Series A - J. Engineering for Power*, vol. 88, No. 1, January 1966, pp 1-12.

2.5   Chauvin, J.: Meridional Flow in Axial Turbomachines. *VKI CN* 99, January 1977.

2.6   Novak, R.A.: Streamline Curvature Computing Procedure for Fluid Flow Problems. *VKI CN* 59, April 1966.

2.7   Novak, R.A.: Streamline Curvature Computing Procedure. *ASME Trans., Series A - J. Engineering for Power*, vol. 89, no. 4, October 1967, pp 478-490.

2.8   Marsh, H.: A Digital Computer Program for the Through Flow Fluid Mechanics in an Arbitrary Turbomachine, Using a Matrix Method. *ARC R&M* 3509, 1968.

2.9   Hirsch, Ch.: Finite Element Method for Through-Flow Calculations. In: *"Through-Flow Calculations in Axial Turbomachinery"*, AGARD CP 195, 1976, paper 5.

2.10   Davis, W.R. & Miller, D.A.J.: A Comparison of the Matrix and Streamline Curvature Methods of Axial Flow Turbomachinery Analysis From a User's Point of View. *ASME Trans., Series A - J. Engineering for Power*, vol. 97, no. 4, October 1975, pp 549-560.

2.11   Renaudin, A & Somm, E.: Quasi Three-Dimensional Flow in a Multistage Turbine Calculation and Experimental Verification. In: *Proceedings of the Symposium on "Flow Reasearch on Blading"*, BBC, CH - Baden, 1969, pp 51-87.

2.12   Cox, H.J.A.: Flow Solution for Axial Turbine Stages. In: *"Transonic Turbines"*, *VKI LS 30*, January 11-15, 1971.

2.13   Cox, H.J.A.: Through-Flow Calculation Procedures for Application to High Speed Large Turbines. In: *"Through-Flow Calculations in Axial Turbomachinery"*, AGARD CP 195, 1976, paper 7.

2.14   Denton, J.D.: Through-Flow Calculations for Transonic Axial Flow Turbines. *ASME Trans., Series A - J. Engineering for Power*, vol. 100, no. 2, April 1978, pp 212-218.

2.15   Novak, R.A.: Flow Field and Performance Map Computation for Axial-Flow Compressors and Turbines. In: *"Modern Prediction Methods for Turbomachine Performance"* AGARD LS 83, June 1976, paper 5.

2.16   Doria, G. & Troilo, M.: Through-Flow Calculations of Large Steam Turbines. In: *Proceedings of the 6th Conference on "Large Steam Turbines"*, Plzen, Czeckoslovakia, May 1979.

2.17   Veuillot, J.P.: Calculation of the Quasi Three-Dimensional Flow in a Turbomachinery Blade Row. *ASME Trans., Series A - J. Engineering for Power*, Vol. 99, No. 1, January 1977, pp 53-62.

2.18   Thiaville, J.M.: Modèles de Calcul de l'Ecoulement dans les Turbomachines Axiales. In: *"Through-Flow Calculations in Axial Turbomachinery"*, AGARD CP 195, 1976, paper 1.

2.19   Horlock, J.G.: On Entropy Production in Adiabatic Flow in Turbomachines. *ASME Trans., Series D - J. Basic Engineering*, vol. 93, no. 4, December 1971, pp 587-593.

2.20   Fruehauf, H.H.: Applicability of Axisymmetric Analysis in Predicting Supersonic Flow Through Annular Cascades. *ASME Trans., Series A - J. Engineering for Power*, vol. 99, no. 1, January 1977, pp 115-120.

2.21   Schröder, H.J. & Schuster, P.: Transonic Flow Through a Turbine Stator Treated as an Axisymmetric Problem. *ASME Paper 73-GT-51*, 1973.

2.22   Through Flow Calculations in Axial Turbomachines. Edited by Ch. Hirsch & J.D. Denton, *AGARD AR 175*, 1981.

2.23   Wolf, H. & Schulz, H.: Untersuchungen zur Berechnung der Meridianströmung in axialen Turbomaschinen. *Wissenschaftliche Berichte der Ingenieur Hochschule Zittau* IHZ-K-79-267, September 1978.

2.24   Katsanis, T.: Use of Arbitrary Quasi-Orthogonals for Calculating Flow Distribution in the Meridional Plane of a Turbomachine. *NASA TN-D* 2546, December 1964.

2.25   Troilo, M.: On the Design of Axial Turbines. In: *Proceedings of the 5th Conference on Fluid Machinery*, vol. 2, Budapest, Akademiai Kiado, 1975, pp 1161-1171.

2.26   Troilo, M.: Criteri di Progetto di Palettature Svergolate di Turbine. *La Termotecnica*, no. 12, Milano, December 1976, pp 618-632.

2.27   Wilkinson, D.H.: Calculation of Blade-to-Blade Flow in Turbomachine by Streamline Curvature. *ARC R&M 3704*, December 1970.

2.28   Wilkinson, D.H.: Stability, Convergence, and Accuracy of Two-Dimensional Streamline Curvature Methods Using Quasi-Orthogonals. *Proceeding of Inst. of Mechanical Engineers*, vol. 184, part 3G(1), 1969-70.

2.29   Troilo, M. & Verta, G.: Aspetti del Flusso Meridiano in Stadi Supersonici di Turbina Assiale. *32° Congress Nazionale ATI*, Rome, September 1977.

2.30   Shaalan, M.R.A. & Daneshyar, H.: A Critical Assessment of Methods of Calculating Slope and Curvature of Streamlines in Fluid Flow Problems. *Heat and Fluid Flow*, vol. 3, no. 1, 1973.

2.31   Nurzia, F. & Troilo: Effetti della Curvatura delle Linee di Corrente sul Flusso Attraverso Turbomacchine Assiali. *Ricerche n. 22, suppl. a La Termotecnica*, Milano, Ago. 1975.

2.32   Craig, H.R.M. & Cox, H.J.A.: Performance Estimation of Axial Flow Turbines. *Proc. Inst. Mechanical Engineers*, vol. 185, 1970-71, pp 407-424.

2.33   Troilo, M. & Rigoli, L.: Previsione del Flusso Meridiano Attraverso uno Stadio Assiale e Confronto Con i Risultati Sperimentali. *37° Congresso Nazionale ATI*, Viareggio, October 1981.

2.34   Lazzati, P. & Macchi, E.: Confronto tra i Risultati Teorici e Sperimentali Relativi ad uno Stadio di Turbina Assiale Monostadio. In *"Metodi di Calcolo della Fluodinamica della Macchine*, ed. CLUP - Milano, 1982.

2.35   Smith, L.H. & Yeh, H.: Sweep and Dihedral Effects in Axial-Flow Turbomachinery. *ASME Trans., Series D - J. Basic Engineering*, vol. 85, no. 3, September 1963, pp 401-416.

2.36   Starken, H.: Transonic Flow in Axial Turbomachinery. In *"Transonic Flows in Axial Turbomachinery"*, VKI LS 84, February 2-6, 1976.

2.37   Dzung, L.S.: Inclined Flow Through an Axial Blade Cascade. *BBC Review*, Vol. 64, No. 6, June 1977, pp 368-

2.38   Traupel, W.: Prediction of Flow Outlet Angle in Blade Rows with Conical Stream Surfaces. *ASME Paper 73-GT-32*, 1973.

2.39   Martelli, F.: Prediction of Subsonic Flow Angle in Turbine Blades. *Fifth World Congress on Theory of Machines and Mechanisms*, Montreal, July 1979.

2.40   Gyarmathy, G.: Grundlagen einer Theorie der Nassdampfturbine. *Dissertation ETH, Zürich, 1962.*

2.41   Gyarmathy, G.: Kondensationsstoss - Diagramme für Wasserdampfströmungen. *Forschung Ing. Wes.*, Bd 29, Heft 4, 1963, pp 105-114.

2.42   The 1967 ASME Steam Tables. New York, ASME.

Three Dimensional Flow Calculation

3.1 CALCULATION OF THE 3D INVISCID FLOW THROUGH A TURBINE BLADE ROW

J.D. Denton

3.1.1 Introduction

Quasi 3D flow calculation methods (e.g. Novak & Hearsay (1977), Katsanis (1966)) have been the main 'tools' of the turbomachinery aerodynamicist for many years. Such methods have been developed to the stage where they are 'robust', flexible and applicable to most types of turbomachines. The quasi 3D approach assumes that the flow is confined to either blade to blade or meridional stream surfaces and the complete solution is obtained by iterating between solutions on these two surfaces. Although this approach inevitably involves some degree of approximation, because some terms in the full 3D equations of motion are neglected, experience shows that good predictions of the flow can be obtained in most cases. Situations in which the quasi 3D methods have difficulty in providing accurate predictions are those involving shear flow, flows with large spanwise components of velocity and geometries with large meridional curvature within a blade row.

Fully 3D solutions are attractive partly because they do not include any approximations and so are potentially more accurate than quasi-3D solutions nor do they have special difficulties with the above types of flow. However, their main attraction is that they are potentially easier to use. This is because it is not necessary to iterate between two separate solutions, a process which invariably involves human intervention and judgement, instead the complete solution is obtained in one calculation without any human intervention. In order to achieve this potential, however, 3D methods have to be developed to the stage where they are as robust, flexible and economical as present Q3D methods. Many of the methods developed to date fail to meet these requirements.

The first fully 3D solutions were reported about a decade ago (Denton (1975), Bosman & Highton (1979), Thompkins (1976)). At first these were too expensive in computer time, too limited in geometry and too temperamental for use in routine design work. However, reductions in computer costs and improvements in numerical methods over the last decade have made 3D methods useable as design tools and they are now used by most large gas and steam turbine manufacturers.

Recent inviscid 3D methods have been described by Laskaris (1978), Hirsch (1982), Soulis (1982), Enselme et al. (1982) and Thompkins (1981). The author's method started its development in 1973 being at first limited to very simple geometries [3.3]. Its capabilities were gradually extended (Denton (1976), Denton & Singh (1979)) with little change in the numerical method, until it could compute the flow through most types of fixed or rotating blade row. During 1981-2 the numerical scheme was changed to use a new grid and algorithm already developed in two dimensions (Denton (1982)). This change proved even more beneficial in 3D than in 2D and enabled rapid development of the method to include even more complex geometries and boundary conditions. The resulting suite of programs is described in this paper.

3.1.2 Basic Method

Inviscid 3D flow calculations are usually based on solutions of either the potential equation or of the Euler equations. There is little doubt that for flows which can reasonably be assumed to be irrotational the potential equation involving only a single variable, (the potential), provides the most economical solutions. However, many turbomachine flows are not irrotational and may involve strong shock waves, in which the potential equation is not useable. The alternative of using the Euler equations involves, in general, solving 5 coupled differential equations and is therefore much less efficient in computer time and computer storage. Despite this handicap the latter approach remains by far the most popular method of solving for 3D flow in turbomachinery. An important contribution to this popularity is its simplicity and ease of understanding and the fact that the same method can be used for all Mach number ranges.

The Euler equations may be solved in either finite difference or in finite volume form. In the former approach the equations are approximated by conventional finite difference equations relating values of the fluid properties stored at grid nodes. In the finite volume method the equations are regarded as conservation equations applied to a series of interconnected elementary volumes. Both approaches are equivalent on a rectangular Cartesian grid. However, for the highly distorted grids which must be used in real turbomachines it is much easier to enforce global conservation (of say mass flow) using the finite volume method. Such conservation is essential in high Mach number flows where a small error in mass flow conservation can lead to large errors in velocity. A further advantage of the finite volume approach is that it is possible to work with a physical rather than a transformed grid and this greatly helps 'Engineering' understanding of the method.

3.1.3 Equations

The equations for conservation of mass, energy and momentum applied to a fixed control volume ΔV for a time interval Δt are simply

$$\Sigma \left(\rho C_x dA_x + \rho C_\theta \, dA_\theta + \rho C_r \, dA_r \right) = \frac{\Delta V}{\Delta t} \cdot \Delta \rho \tag{3.1.1}$$

$$\Sigma \left(\rho C_x h_t \, dA_x + \rho C_\theta h_t dA_\theta + \rho C_r h_t \, dA_r \right) = \frac{\Delta V}{\Delta t} \cdot \Delta(\rho E) \tag{3.1.2}$$

$$\Sigma \left((P + \rho C_x^2) dA_x + \rho C_\theta C_x dA_\theta + \rho C_r C_x dA_r \right) = \frac{\Delta V}{\Delta t} \cdot \Delta(\rho C_x) \tag{3.1.3}$$

$$\Sigma \left(\rho C_x C_r \, dA_x + \rho C_\theta \, C_r \, dA_\theta + (P + \rho C_r^2) \, dA_r \right) + \rho \Delta V \cdot \frac{C_\theta^2}{r} + P \cdot \frac{\Delta V}{r}$$

$$= \frac{\Delta V}{\Delta t} \cdot \Delta(\rho C_r) \tag{3.1.4}$$

$$\Sigma \left(\rho C_x rC_\theta \, dA_x + (P + \rho C_\theta^2) \quad r \, dA_\theta + \rho C_r rC_\theta \, dA_r \right) = \frac{\Delta V}{\Delta t} \cdot \Delta(\rho rC_\theta) \tag{3.1.5}$$

The summations are taken over the faces of the volume and dA_x, dA_r, dA_θ, the projected areas of these faces in the coordinate directions, are taken as positive when directed inwards to the element.

In the above equations C_x, C_θ, C_r are the absolute velocities and x, r, θ are cylindrical coordinates, i.e., the radial and circumferential directions are always defined relative to the point at which the velocity is being evaluated and so vary around the faces of an element. Equation 3.1.5 is the moment of momentum equation which is used because it is simpler than the θ momentum equation. Note that in 3D flow it is seldom possible to replace the energy equation (3.1.2) by an assumption of constant stagnation enthalpy as is usually done in 2D flow. Hence the present programs always solve the energy equation.

Equations 3.1.1-3.1.5 can all be expressed in the form :

$$\Sigma \; \text{FLUX} + \text{SOURCE} = \frac{\Delta V}{\Delta t} \cdot \Delta \; (\text{property}) \tag{3.1.6}$$

where the SOURCE term only occurs in the radial momentum equation and FLUX is the rate of flow of mass, energy etc., through each face of the element. For use in rotating blade rows the equations could be re-cast in terms of relative velocities, with the inclusion of extra terms where necessary, and the problem solved in a rotating coordinate system. In practice it has been found simpler to continue to work with absolute velocities and to transform the rate of change calculated from equation 3.1.1 to 3.1.5 for a fixed grid to a rotating grid which instantaneously coincides with the fixed grid, using

$$\left.\frac{\partial Y}{\partial t}\right)_R = \left.\frac{\partial Y}{\partial t}\right)_F + \Omega \cdot \frac{\partial Y}{\partial \theta}$$

where Y is ρ, ρE, ρC_x, $\rho r C_\theta$ or ρC_r depending on which equation is being updated. Integrating this equation over the elemental volume shows that the $\Omega \frac{\partial Y}{\partial \theta}$ term can be expressed simply as an additional flux through the bladewise faces of the elements.

3.1.4   Grid

The choice of grid is probably the most critical part of any 3D flow calculation procedure. This is especially true in turbomachinery where extremely complex geometries have to be handled. The use of as simple a grid as possible greatly helps in fitting these complex geometries but a simple grid is unlikely to provide the greatest accuracy from a limited number of grid points. In the author's original method the finite volumes consisted of cuboids with a node at the center. In two dimensions [3.13] it was found to be simpler and more accurate to use essentially the same cuboids but with a node at each corner. This new type of grid has been extended to 3D where its advantages over the old grid have proved even greater than in 2D. The cuboids are formed by the intersection of 3 families of surfaces as shown in figures 3.1.1 and 3.1.2. The streamwise surfaces are akin to the blade to blade stream surfaces of the Q3D approach. They are surfaces of revolution such that the first surface coincides with the hub of the machine and the last with its shroud or casing. The remainder of the streamwise surfaces are non-uniformly spaced between the hub and the shroud with the actual spacing being chosen by the user.

The bladewise surfaces are such that the first and last of them contain the two blade surfaces forming one blade-blade passage. Upstream and downstream of the blade row these two surfaces are chosen to lie roughly parallel to the relative flow direction and they are one blade pitch apart in the circumferential direction. The remainder of the bladewise surfaces are non-uniformly spaced, as chosen by the user, between the two blade surfaces.

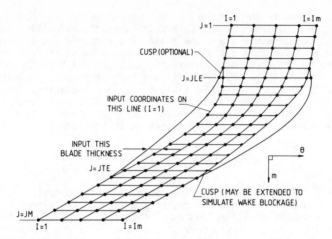

FIGURE 3.1.1 Grid on a streamwise (i.e., m-θ) surface
(grid is usually much finer than shown above)

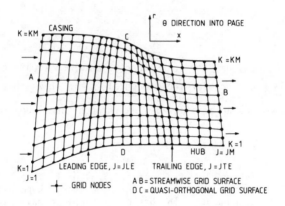

FIGURE 3.1.2 Grid in a meridional plane

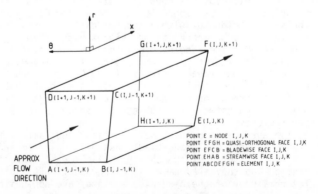

FIGURE 3.1.3 Single element of grid

The quasi-orthogonal surfaces are surfaces of revolution lying roughly perpendicular to the meridional flow direction. One of these surfaces contains the leading edge of the blade row and another the trailing edge, the positions of the remainder of the quasi-orthogonal surfaces may be chosen by the user. There are grid nodes, where all flow properties are stored, at each of the eight corners of the cuboid. However, except on the boundaries, each corner is shared by eight cuboids so there is effectively only one grid node per cuboid.

The projected area of every face of the cuboid in the three coordinate directions must be evaluated and stored. Since the streamwise and quasi-orthogonal faces lie in the circumferential direction this makes a total of seven areas per element. The areas are projected in coordinate directions defined at the center of the face concerned.

3.1.5 Solution Procedure

The numerical scheme used to solve equations 3.1.1-3.1.5 is that described as scheme 'A' by Denton (1982). The other two schemes (B & C) described in that paper are not stable when the energy equation is included. Hence the scheme used is the exact analog of the opposed difference scheme used in the author's original methods, but is applied to the new grid.

In each equation the fluxes through all the faces of the elements are evaluated using averages of the flow properties at the four corners of the face concerned. These fluxes are summed to find the change of the conserved property for each element over the time step. One quarter of this change is then added to the values of the property at the four downstream corners of the element. Note that the manner of distributing the changes does not affect the final steady state solution, for which the sum of the fluxes must be zero, but it does have a critical influence on the stability of the method.

Stability is enforced by using an effective pressure in the three momentum equations rather than the true pressure. This effective pressure is made equal to the current pressure at the next downstream grid point plus a correction, i.e.,

$$P_{EFF_J} = P_{J+1} + CF_J \qquad (3.1.7)$$

The correction CF is obtained by an interpolation which must not make use of the true pressure at J. Changes in CF are damped after every time step according to

$$CF_{J,NEW} = (1-RF) \times CF_{J,OLD} + RF \times CF_{J,INT}$$

RF is a relaxation factor, typically 0.05. The current scheme uses parabolic interpolation, as illustrated in figure 3.1.4, and so provides second order accuracy for a continuously changing pressure. Additional artificial viscosity may be provided in shock waves by taking only part of the interpolated correction, the fraction used being a function of the local density gradient so that outside the shock region it is very closely one.

Some small amount of smoothing in the circumferential and quasi-orthogonal directions is necessary to stabilize the scheme. As explained in [3.13] this is because there are more nodes than elements and the smoothing is needed to remove this indeterminacy in the problem. The amounts of smoothing needed are very small indeed. Typically during every time step all independent variables, such as ρ, are smoothed according to

$$\rho_{NEW} = (1 - SF) \times \rho_{OLD} + SF \times \rho_{SMOOTHED} \qquad (3.1.8)$$

where SF is of the order of 0.005 and $\rho_{SMOOTHED}$ may be obtained by either linear or non-linear smoothing. The latter is formulated so that it does not change a cubic variation in the variable (e.g. ρ) and so introduces only extremely small

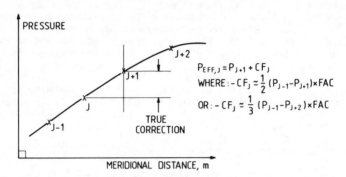

FIGURE 3.1.4   Definition of correction factors

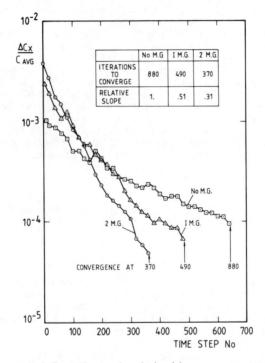

FIGURE 3.1.5   Effect of multigrid on convergence

amounts of artificial viscosity. Numerical experiments show that the smoothing
factor SF may often be increased by about an order of magnitude before it starts
to significantly affect the solution.
   The scheme as described so far is only stable if the flow velocity is positive in the meridional direction. This is a feature of the opposed difference
scheme which was overcome in two dimensions by means of a modified scheme (scheme
B of [3.13]) applied everywhere in the flow field. The 3D programs overcome the
problem by switching to the use of upwinded values of convected velocity (but
not mass flux, pressure or enthalpy) wherever the meridional velocity is detected
to be negative. This permits large regions of reversed flow to be predicted at

the expense of a loss of second order accuracy in such regions.

3.1.6 Boundary Conditions

The boundary conditions used are simple and intuitively obvious. The only condition applied on solid boundaries (blade and endwalls) is no flow through the surface. The flow velocity is not resolved parallel to the surface since it is simpler not to do so and not obvious that the errors incurred (i.e., incorrect acceleration perpendicular to the surface) are worse than those caused by resolving (i.e., body force imposed perpendicular to the surface).

A periodicity condition is imposed on the bladewise surfaces upstream of the leading edge and downstream of the trailing edge and points on these surfaces are effectively updated exactly as interior nodes.

At the downstream flow boundary only the static pressure needs to be prescribed. This may either be input and held constant at all spanwise positions or only the value on the hub may be fixed and the pressure at other points built up from an assumption of simple radial equilibrium as the calculation proceeds. The latter assumption may not be realistic if the downstream boundary coincides with a region of high meridional curvature.

At the upstream (inflow) boundary the stagnation pressure and stagnation temperature must be specified at each grid node, the values may vary in both the spanwise and circumferential directions. Several options are available for specifying the inlet flow directions. Either the relative or the absolute flow directions in both the streamwise and meridional planes may be specified, or the angle in the streamwise plane may be replaced by a specified absolute whirl velocity. The latter condition can be used when the relative inlet velocity is supersonic and the flow will then automatically satisfy the unique incidence condition.

The final condition needed at the upstream boundary is a means of extrapolating the static pressure from the interior flow field. This may be done by applying a condition of either $dp/ds = 0$ or $d^2p/ds^2 = 0$ along the quasi-streamlines. The pressure so obtained is used together with the stagnation pressure and temperature to obtain the absolute velocity. The flow directions or specified whirl velocity then enable this to be converted into the three velocity components.

3.1.7 Stability and Convergence

As with all explicit schemes the maximum stable time step is limited by the CFL condition to

$$\Delta t < \frac{\Delta \ell}{C+a} \qquad (3.1.9)$$

In complex geometries with highly distorted elements it is difficult to know which values of $\Delta \ell$ and C limit stability. In practice $\Delta \ell$ is taken as min (Δs, $\Delta(r\theta)$, Δq) for each element and a factor of safety is provided by taking a as the inlet stagnation speed of sound and C as the estimated maximum velocity in the flow field.

Either uniform time steps (which are the same for all elements and equal to the smallest time step needed by any one element), or non-uniform time steps, may be used. The latter option takes different time steps for each element, based on the local stability, and usually gives much more rapid convergence. It also enables the time step to be varied in response to the rate of change of the variable in the element concerned. If the rate of change is found to be large the time step is automatically reduced so that only a small increment is added to the variable. This 'negative feedback' can be made to exert a powerful stabilizing influence and enables acceptable solutions to be obtained over most of the flowfield even when local regions are unstable. It also prevents the time step having to be reduced in order to overcome initial transients.

The rate of convergence may also be greatly accelerated by the use of a multigrid type of method. This is described in [3.13] for 2D flow and carries over directly to 3D. The method consists simply of combining groups of elements into blocks and updating each node by an increment from its adjacent elements and, in the same cycle, by an increment from the adjacent blocks. In the 3D codes two levels of multigrid have been coded and an advantage of the method used is that the second level uses very little extra computer work per cycle. Typically blocks of $3 \times 3 \times 3$ and $9 \times 9 \times 9$ elements are used and the resulting effect on convergence is shown in figure 5. In interpreting this graph it should be remembered that when multigrid is used the same change in a variable implies a much smaller (1/3 or 1/9) flux imbalance per element than without multigrid. Hence the factor of 3 increase in convergence rate shown in figure 3.1.5 is an underestimate of the rate of reduction in flux imbalance.

Convergence is usually taken to occur when the average change in meridional velocity per cycle is less than $5 \times 10^{-5} \times$ an average velocity for the whole flow. The actual number of cycles to convergence depends on the number of grid points used. Values for some of the test cases used in this paper are given in table 3.1.1. Convergence in 300-500 cycles is usual for many practical problems.

Computer time per point per cycle clearly depends on the machine used. All the results shown were obtained on a Perkin Elmer 3230 mini computer for which the time per grid point per cycle was about 1.3×10^{-3} s. Times on a modern mainframe computer would be about 1/10 of this. Storage requirements are about 1 M byte per 9000 grid points.

Blade row	Grid used No. of points	Iterations to converge	Total CPU time* minutes
3D transonic turbine cascade Fig. 3.1.6	$10 \times 49 \times 10$ 4900	300	32
Swept low speed turbine cascade Fig. 3.1.8	$10 \times 60 \times 10$ 6000	370	48
3D steam turbine stage Fig. 3.1.10	$10 \times 89 \times 13$ 11570	860	280

*Times on Perkin-Elmer 3230 mini-computer

TABLE 3.1.1

3.1.8 Examples of the Use of the Basic Program

The basic version of the program can be used for all types of blade row which do not contain splitter blades or shrouds. The only other geometrical restriction is that the hub and casing must be surfaces of revolution. All these limitations have been separately removed in special versions of the code.

A comparison of the present method with the author's original method and with test data is shown in figure 3.1.6. This shows results for the 3D transonic turbine cascade tested by Camus et al. (1983). Both methods give similar results over the upstream half of the blade with the new method on the whole being rather better. In particular the methods give almost identical results on the pressure surface at 8.3% span which suggests that the discrepancy from experiment in this region may be due to viscous effects. Past experience has shown that the shock system from the trailing edge of a transonic turbine blade cannot be computed

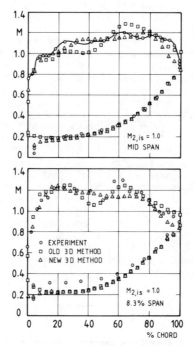

FIGURE 3.1.6 Comparison of old and new methods with experiment
on a 3D transonic turbine cascade

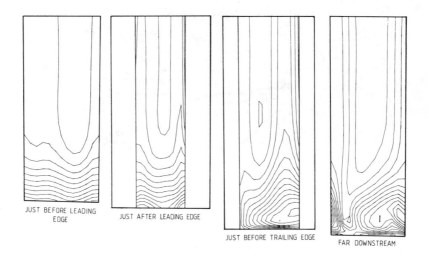

FIGURE 3.1.7 Stagnation pressure contours in transonic turbine
cascade with shear flow

without an extremely fine grid and special treatment of the trailing edge region.
No such treatment was included in either method and so neither predicts the flow
over the rear of the suction surface with great accuracy. The trends and levels
of Mach number are, however, correctly predicted by both programs.

As an illustration of the use of the program to predict shear flows the
same test case has been run with a stagnation pressure gradient imposed at the
upstream boundary. Accurate prediction of such inviscid secondary flows requires
a very fine grid and minimal numerical viscosity. For the present case 19 grid
points were used in the spanwise direction with their spacing varied so as to
concentrate more points in the simulated boundary layer which was imposed only
on the flared endwall. Figure 3.1.7 shows contours of stagnation pressure on
quasi-orthogonal surfaces through the blade row and illustrates typical features
of inviscid secondary flow. Similar results have been obtained by Barber (1981)
using the author's original program.

Swept blade rows provide an extremely simple 3D geometry which nevertheless
produces complex 3D flows which are difficult to predict from a quasi-3D approach.
Experimental data on such a turbine cascade has been obtained by Gotthardt & Stark
and results are available in [3.16]. Figure 3.1.8 compares experimental and
computed results for the blade surface pressure distributions close to the endwalls. The agreement for this low speed flow can be seen to be very good. Comparisons of the predictions of the original 3D program with test data for similar
simple 3D geometries have been published by Dawes and Squire (1982).

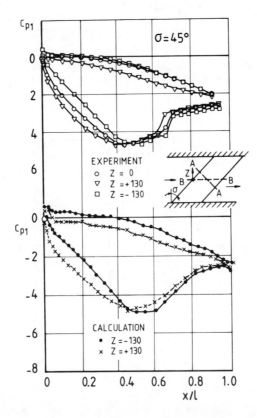

FIGURE 3.1.8 Swept turbine cascade

3.1.9   Complete Stage Calculations

The flow through two blade rows in relative rotation is inherently unsteady unless they are widely spaced. However, except when specifically studying unsteady effects it is usual to completely neglect the circumferential variations in flow at entry to a blade row and to assume that the interaction between two rows is confined to a spanwise variation in flow properties. This spanwise variation would usually be obtained from a throughflow type of calculation in the meridional plane and the output from such a calculation would be used as a circumferentially uniform inflow to a Q3D or 3D blade-blade calculation.

If similar assumptions are made in a 3D calculation it is possible to predict the flow through two or more blade rows in relative rotation in a single calculation. All that is required is that each blade row should be presented with a circumferentially uniform inflow. Hence it is necessary to circumferentially average all flow properties at some point between each pair of blade rows. Since all fluxes are conserved between the averaged flow and the non-uniform flow upstream of it this averaging process does not introduce any source of mass, momentum or energy into the flow. However, the entropy of the averaged flow will not exactly equal the mass averaged entropy of the upstream flow and this is a price which must be paid for the ability to perform such calculations. Numerical experiments show that the change in entropy is small even for highly non-uniform flow.

The grid used for stage calculations is illustrated in figure 3.1.9. Since, in general, there will be different numbers of blades in each row the grids will be discontinuous at the mixing plane and the discontinuity in area of the quasi orthogonal faces of the elements must be allowed for when satisying the conservation equations. The only limit to the number of blade rows which can be calculated in this way is provided by computer storage limitations. The spanwise variation of the flow properties at the mixing plane is not specified but is a product of the calculation exactly as it is in throughflow calculations. This constitutes one of the main attractions of the procedure since it enables a stage to be designed using a single calculation method rather than by iterating between throughflow and Q3D or 3D blade-blade methods.

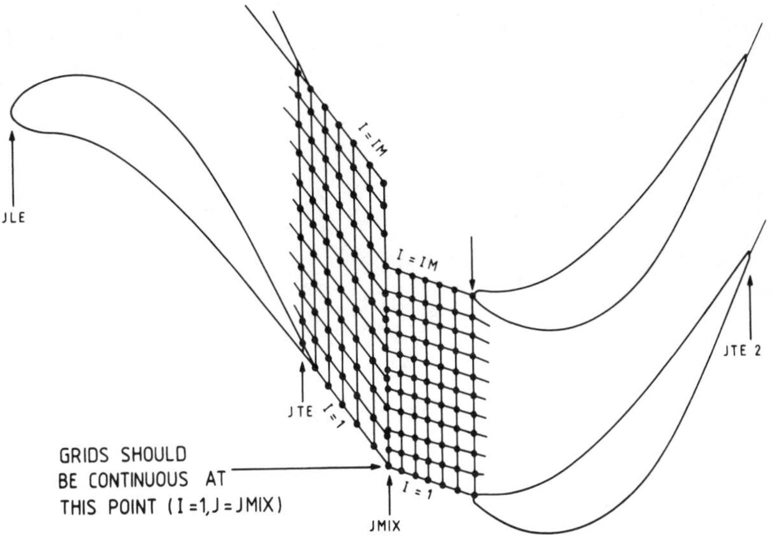

FIGURE 3.1.9 Detail of grid junction for stage calculation

An example of the predictions for the last stage of a large steam turbine is shown in figures 3.1.10 - 3.1.13. This highly 3D flow with Mach numbers ranging

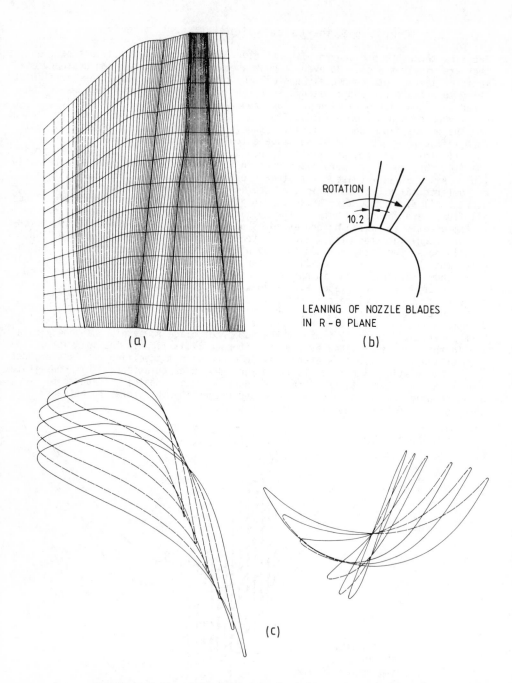

FIGURE 3.1.10   ANSALDO hypothetical last stage design

from 0.1 to 1.8 presented no special difficulty for the program. A maximum of
about 13000 grid points could be used and these were distributed as 89 in the

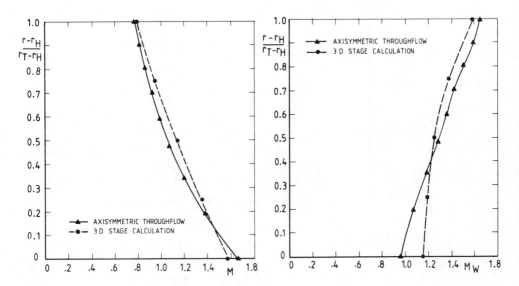

FIGURE 3.1.11 Nozzle and rotor exit Mach numbers
for ANSALDO hypothetical last stage design

axial direction, 13 evenly spaced in the spanwise direction and 10 evenly spaced
in the pitchwise direction. A meridional view of the grid is shown in figure
3.1.10a.

In order to approximate the properties of equilibrium wet steam by a perfect
gas it is necessary to match real wet steam properties by the following equations:

$P = \rho RT$ R = gas constant

$h_t = CpT$ Cp = specific heat capacity

$PT^{\frac{\gamma}{\gamma-1}} = $ constant γ = ratio of specific heats

    $= \left(Cp/(Cp-R) \right)$

This is best done for low pressure wet steam by choosing

R = 436.5 J/kg/K

Cp = 7365 J/kg/K

Hence γ = 1.063

The calculation converged in 830 iterations. The mass flow predicted was
53.9 kg/s, compared to a mass flow of 56.7 kg/s obtained with an axisymmetric
throughflow calculation. One would expect the throughflow calculation to give
less flow than the inviscid 3D calculation since it includes losses. Figure
3.1.11 compares the 3D and throughflow predictions for Mach number distribution
after the nozzle and rotor. The differences are significant but not large.

Figure 3.1.12 shows blade surface Mach number distributions at root, mid-
height and tip. The discontinuity in the velocity distribution at the mixing

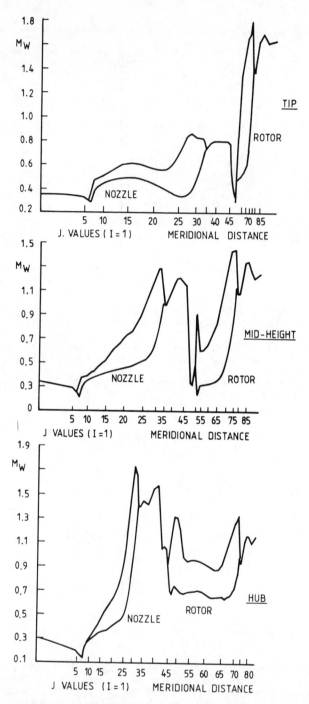

FIGURE 3.1.12 Blade velocity distributions
for ANSALDO hypothetical last stage design

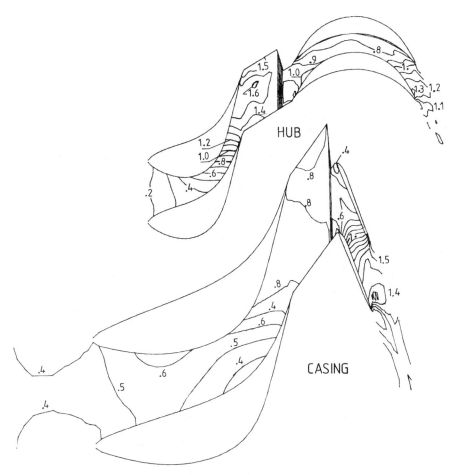

FIGURE 3.1.13 Mach number contours for
ANSALDO hypothetical last stage design

plane is due to the change from absolute to relative Mach numbers for the rotating blade row. As is often found there is a problem at the rotor root where the
relative inlet flow is about sonic and the suction surface flow accelerates to
$M \cong 1.3$ before shocking down at about 25% chord. At mid-height there is again a
suction peak at the leading edge of the rotor. This is a local effect which
could easily be cured by minor changes to the leading edge geometry. At the tip
there is some diffusion on both the nozzle blade surfaces due to the high flare
angle and some thickening of the stator at around mid chord would be desirable.
The rotor tip is at a good incidence and shows highly accelerating flow throughout. The whole rotor appears to be at about its limit load with the shock at
the trailing edge. Figure 3.1.13 shows the Mach number contours in the blade-
blade plane at hub and tip.
     In order to illustrate the effects of various geometrical features a series
of modifications were made to the nozzle geometry with no change to the rotor or
to the thermodynamic properties. These changes were not necessarily intended to
benefit this machine but were to illustrate the trends and magnitude of the effects
produced by features which might be used either in retrofitting new nozzles to an
existing machine or in developing a new design.

The modifications tried were :
(a) Twisting the nozzle blade so that its opening was increased at the root and
slightly decreased at the tip. The actual changes were made by the computer and
were done by shearing the blade sections rather than rotating them. As a result
it is not possible to say exactly how large the changes in opening:pitch ratio
were. However, the suction surface angle near the trailing edge was decreased
(from axial) by about 4° at the hub and increased by about 1° at the tip.
(b) Leaning the nozzle from the radial direction by 15° in addition to its original lean. The direction of lean being such that the pressure surface is inclined
inwards. This produces a radially inwards component of blade force on the fluid.
(c) A combination of (a) + (b).
      The results of these modifications are presented as the Mach number distribu-
tion at nozzle exit in Fig. 3.1.14. The objective in all cases was to reduce the

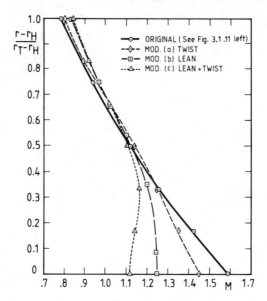

FIGURE 3.1.14 Effect of lean and twist on nozzle exit
Mach number of ANSALDO hypothetical last stage design

gradient of this Mach number so as to facilitate design of a following rotor. In
particular to increase rotor root reaction.
      All modifications produce some reduction in the root Mach number but the ef-
fects of lean (b) are by far the greatest of any individual change whilst those
of the twist (a) are the least. The combination of twist with lean (c) produces
a large increase in root reaction and induces too much negative incidence on the
present rotor blade.
      The effects at the rotor tip are comparatively small in all cases. It is de-
sirable to increase the nozzle exit Mach number here so as to reduce the rotor inlet velocity. However, a negative feedback situation exists whereby increases in
mid-height Mach number cause an increase in radial pressure gradient which acts
to reduce the tip Mach number. Variations in casing geometry in the tip region
would probably be more effective but have not been tried.

3.1.10   Conclusions

Fully 3D flow calculations have evolved to the stage where they can be used efficiently in the design process of last stage steam turbine blade rows. The key

feature of the present method is the simplicity of the grid and of the numerical scheme. Although the grid is certainly not the most efficient in terms of accuracy from a limited number of points it would be difficult to cover a wide range of geometries with a more complex grid. The simplicity of the numerical scheme makes it very fast in terms of CPU time per grid point per time step and, perhaps more important, enables the code to be readily understood and developed by other users. The human effort involved is much less than when using iterative through-flow-blade to blade methods. A solution requires typically 1 hour human time and 4 hours computer time to perform, plot and run the data.

The weakest point of the code, and of most other methods of solving the Euler equations in turbomachinery, is the failure to accurately capture the trailing edge shock system from transonic turbine blades. Work on this and on other developments is continuing and it is anticipated that such improvements will be progressively incorporated into the existing infrastructure of the code.

ACKNOWLEDGEMENTS

The development of these codes has benefitted enormously from the feedback obtained from many users of this and of earlier 3D codes. The author would like to express his gratitude to all such users and to encourage them to keep him informed of both the successes and failures of the method.

3.2 DEVELOPMENTS IN THE CALCULATION OF 3D VISCOUS COMPRESSIBLE FLOW IN AXIAL TURBOMACHINERY

W.N. Dawes

3.2.1 Introduction

Recent years have seen intensive development of computational fluid dynamics and increasing application to problems in turbomachinery. The reasons for this are two-fold. Firstly, the ever increasing cost of high quality experimentation means that it is very cost-effective to study a variety of configurations and builds by numerical simulation and select just a few promising designs for rig testing. Secondly, the numerical simulation of the flow field can be used to provide real insight and understanding leading to improved design concepts. Reference [3.18] discusses typical uses of computational fluid dynamics in the aero engine design environment.

Three dimensional inviscid flow simulations, using either potential flow solvers or Euler solvers, are now routine and widely used by most engine manufacturers. An extensive review of computational methods in turbomachinery is given in [3.19]. Attention is now directed towards dealing with the range of blade row features illustrated in figure 3.2.1 (taken from [3.19]). These real flow features include secondary flows, tip clearance flows, blade, hub and casing boundary layers, their interaction with each other and with three dimensional shock systems, and so on. These phenomena are complicated and may dominate particular aspects of the performance of the machine, for example the stall margin of an axial flow compressor.

A wide-ranging review of 3D computational techniques applied to internal flows in propulsion systems is given in [3.20]. For reasons of space this review article will be limited to axial-flow turbomachinery and to computational techniques for solving the fully 3D Navier-Stokes equations. It is considered that only this level of sophistication is capable of dealing with the real flow features mentioned above.

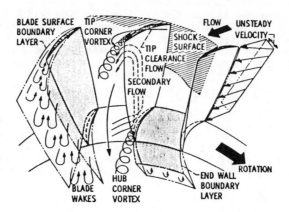

FIGURE 3.2.1   Real flow features in an axial turbomachine
              from McNally & Sockol [3.19]

3.2.2 3D Reynolds Averaged Navier-Stokes Equations

The unsteady Reynolds averaged Navier-Stokes equations in a stationary frame may
be written as [3.21].

$$\frac{\partial \rho}{\partial t} + \nabla \cdot \rho \bar{u} = 0$$

$$\frac{\partial \rho \bar{u}}{\partial t} + \nabla \cdot (\rho \bar{u}\bar{u} - \bar{\bar{\tau}}) - \rho \bar{f} = 0 \qquad\qquad (3.2.1)$$

$$\frac{\partial \rho E}{\partial t} + \nabla \cdot (\rho E \bar{u} - \bar{u} \cdot \bar{\bar{\tau}} + \bar{q}) - \rho(\bar{f} \cdot \bar{u}) = 0$$

where $\bar{u} = (u,v,w)$, $\bar{\bar{\tau}}$ is the stress tensor, $\bar{f}$ a body force. The set of equations
must be closed by an equation of state and a turbulence model. Currently, alge-
braic eddy viscosity models are the most common way of simulating turbulence; a
recent review of turbulence modelling is given in [3.22]. In a rotating frame
of reference, additional terms must be added to equation (3.2.1).
    There are two main approaches to the solution of the equations of motion :
parabolic methods and elliptic methods.

3.2.3 Parabolic Methods

In the parabolic method, a primary flow direction is assumed and solution is by
spatial marching. The pressure is usually considered to be elliptic and stored
throughout the flow field, but all the other variables are assumed to be parabolic and are stored just in two or three cross flow planes. The equations of
motion are simplified by neglecting streamwise diffusion terms and flow separation is not permitted.
    The momentum and energy equations are solved in the presence of the assumed
pressure field by marching downstream from plane to plane; local continuity is
forced by adjusting the flow variables. Then an elliptic equation is solved
over the whole flow to update the pressure field, global mass conservation is
forced. The whole process can be repeated until convergence ("partially parabolic method") or the pressure itself can also be considered parabolic and the
solution terminated after one sweep ("fully parabolic method"). References [3.23]

to [3.27] contain extensive information concerning parabolic methods; no more
details will be given here.

The main benefits of the parabolic approximations are the small computer
memory requirements and low CPU times. These are paid for by the restricted range
of flows which can be dealt with. The method is best suited to duct-like flows
with moderate turning. Relatively strong secondary flows can be handled provided
there is no flow reversal in the primary flow direction. The application of
parabolic methods to turbomachinery has been mainly restricted to centrifugal
devices (Refs. [3.28] to [3.30] for example) and associated flow problems [3.31].

3.2.4   Elliptic Methods

Methods which permit all the flow variables an elliptic variation are the main
interest of this review article. Solution methods differ in their choice of
either the steady or time unsteady form of the equations of motions as their
starting point.

3.2.4.1   *Steady Flow Methods*

The SIMPLE algorithm [3.32] has been a popular way of solving the steady flow
equations. Hah, in particular, has published a number of pioneering computations
using his adaptation of the algorithm ([3.33] to [3.35]). The governing equations
along with suitable turbulence closure equations, are expressed in the general
form :

$$(\rho U_i \phi)_{,i} = (\Gamma \phi_{,i})_{,i} + S_\phi \tag{3.2.2}$$

where ϕ is any of the dependent variables (u,v,w,k or ϵ), Γ is the diffusion coefficient and S_ϕ are the remaining source terms. Finite difference equations are
obtained by integrating the differential equations over arbitrarily shaped control volumes. The mean velocities, temperature and turbulence quantities are
solved for with an assumed pressure field. The pressure field is adjusted by
combining a velocity-pressure correction equation with the continuity equation,
and global mass conservation satisfied. The corrected pressure field is then
used again with the transport equations for momentum etc. and the whole process
repeated. The overall procedure is iterative and resembles classical relaxation
methods (SOR, ADI, etc.) in the way the variables are updated.

The problem with the method is that iterative stability cannot be guaranteed
if local cell Reynolds numbers are too high. It is usual to switch the effective
discretisation of the convection terms from centered to upwinded if local cell
Reynolds numbers exceed two. This introduces significant "numerical viscosity"
which may severely compromise accuracy. Hah uses both "quadratic upstream differencing" and "skew upwind differencing" to control this as far as possible.

Hah has obtained impressive results [3.33] for the secondary flow development (and leading edge horseshoe vortex structure) in the linear cascade of turbine blades reported by Langston [3.36], for the flow in a low speed compressor
rotor [3.34], and for the tip leakage flow in low speed axial compressor rotor
[3.35]. Results from the last of these applications are reproduced in figure
3.2.2.

Moore & Moore [3.37] have presented an elliptic code solving the steady
flow equations. They use a pressure-correction solution procedure which has
similarities with Hah's method and other SIMPLE-based methods. In particular,
the governing equations are discretized on a general finite volume mesh and the
momentum transport equations solved with a fixed pressure field. The pressure
field is then adjusted using a pressure-correction equation devised to force
global conservation.

Measurement by Hunter and Cumpsty

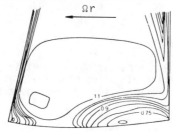

Prediction

a)   Flow coefficient φ = 0.7

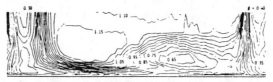

Measurement by Hunter and Cumpsty

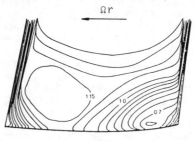

Prediction

b)   Flow coefficient φ = 0.4

FIGURE 3.2.2   Tip clearance flow in a low speed axial compressor rotor;
comparison of rotor exit relative dynamic head, from Hah [3.35]

However, in response to the need to ensure iterative stability (which in
this context means ensuring a diagonally dominant system) Moore & Moore have
adopted a novel approach. Rather than using upwind differencing to represent
convective fluxes when cell Reynolds numbers are high, they employ a technique
they call "upwinded control volumes". When the momentum equations are integrated
over a particular control volume, the net flux imbalance produces a change in
momentum, Δρu for example. This "change" is then assigned to a point in such a
way that the momentum control volume is on the upstream side of the grid point.
This technique is elaborate and complicated, especially when the flow direction
is reversing through a separation bubble for example, but has the advantage of
stabilizing the finite difference equations without the need to introduce

spurious artificial viscosity terms. Once the iterations have converged, the "change" is zero and so its assignment is immaterial.

Moore & Moore have reported good results for the secondary flow development [3.37] in a linear cascade of turbine blades. Figure 3.2.3 reproduces some of their results showing the development of the passage vortex : velocity vectors near the end wall and total pressure loss contours at several axial stations.

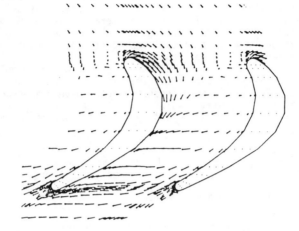

a)   Velocity vectors near the endwall

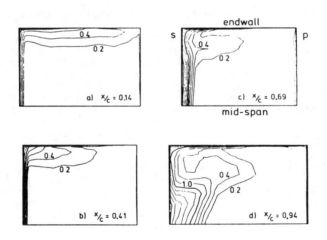

b)   Total pressure loss contours

FIGURE 3.2.3   Development of the passage vortex in a low speed linear cascade, from Moore & Moore [3.37]

3.2.4.2 Time Marching Methods

Time marching methods for solving the unsteady 3D Navier-Stokes equations have
been the object of intensive research for external flow simulations [3.38] but
have only recently started to be developed for application to turbomachinery geometries. Retaining the time derivatives in the governing equations has the advantage that the equations are parabolic/hyperbolic in time regardless of whether
the flow is locally subsonic or supersonic; shock waves may be captured automatically using a suitable artificial viscosity term. This is in contrast to the
steady flow methods discussed before which would not be able to treat transonic
shocked flow without some sort of type-dependent differencing procedure, like that
employed with potential flow solvers.

At high Reynolds number, explicit schemes such as the popular MacCormack
algorithm [3.39] have very restrictive stability properties arising from the CFL
time step limit and the fine mesh necessary to resolve blade boundary layers.
We can, at least in principle, resolve this stability problem by using an implicit
algorithm like the Beam-Warming or Briley-McDonald schemes [3.40, 3.41]. For
the model equation

$$\frac{\partial u}{\partial t} + \frac{A\partial u}{\partial x} = \frac{\nu \partial^2 u}{\partial x^2} \tag{3.2.3}$$

such an implicit algorithm takes the form

$$\left[I + \Delta t \left(A\delta_x^0 - \nu\delta_{xx}^2 \right) \right] \Delta u = - \Delta t \left(A\delta_x^0 u - \nu\delta_{xx}^2 u \right) = R \tag{3.2.4}$$

where δ_x^0 and δ_{xx}^2 are centered difference replacements for first and second
derivatives and $\Delta u = u^{n+1} - u^n$ with n the time level. R represents the residue
and R = 0 is the required result. Weinberg et al. [3.42] have presented 3D
Navier-Stokes solutions for the transonic flow in a linear cascade of gas turbine
blades. They use the differential equations of motion transformed to a general
curvilinear co-ordinate system and expressed in "semi-strong" conservation form.
In three dimensions the implicit operator in equation (3.2.4) is factored into
three narrow block-banded matrices associated with the three coordinate directions. This block ADI technique reduces the multidimensional problem to three
efficiently solved one dimensional problems

$$\left[L_I \right] \left[L_J \right] \left[L_K \right] \Delta u = R_{3D} \tag{3.2.5}$$

Figure 3.2.4 shows predicted velocity vectors at two spanwise stations in the
linear cascade [3.42]. The presence of a saddle point just near the end wall
associated with the leading edge horseshoe vortex structure is clearly indicated.

The residue in equation (3.2.5), R_{3D}, is formed using a finite difference
approximation centered in space. The use of a centered approximation to represent the convection terms is desirable from the points of view of accuracy, spatial symmetry, compactness and simplicity. However, in convection dominated flow,
centered space differences lead to two main numerical difficulties. The first
is the need to add artificial dissipation to prevent wiggles (i.e., odd-even
point solution decoupling) and to permit shock capture. The second difficulty
is that although the implicit algorithm, equation (3.2.4), in theory has very
good stability properties, in practice the ability to invert the left hand side
matrix is compromized by the loss of diagonal dominance associated with large
time steps - and large time steps must be used to pay for the increased overhead.
In fact, the faithful mimicry of the right hand side in the linearized left hand
side matrix perpetuates the basic problem of centered difference convection
terms - lack of diagonal dominance.

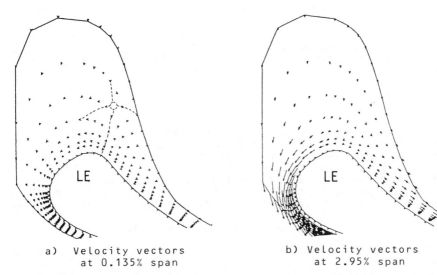

a) Velocity vectors
at 0.135% span

b) Velocity vectors
at 2.95% span

FIGURE 3.2.4 Secondary flow development in a transonic linear cascade,
from Weinberg et al. [3.42]

The present author has developed an implicit algorithm based on the idea of premultiplying the basic implicit algorithm (3.2.4) by a suitable pre-processing matrix [3.43] . If we premultiply by $\left[I - \Delta t \left(A\delta_x^0 - \nu\delta_{xx}^2\right)\right]$ we get

$$\left[I - \Delta t \left(A\delta_x^0 - \nu\delta_{xx}^2\right)\right] \left[I + \Delta t \left(A\delta_x^0 - \nu\delta_{xx}^2\right)\right] \Delta u = R^* \qquad (3.2.6)$$

where $R^* = \left[I - \Delta t \left(A\delta_x^0 - \nu\delta_{xx}^2\right)\right] R$

The object of the present work is to obtain steady solutions, (i.e., $\Delta u \to 0$) and so we regard the left hand matrix as merely providing time marching stability. Accordingly, we multiply out the left hand side of (3.2.6) and neglect terms which cannot be supported on a three point molecule so that we retain the advantages of a tridiagonal structure,

$$\left[I - \varepsilon\Delta t^2 A^2 \delta_{xx}^2\right] \Delta u = R^* \qquad (3.2.7)$$

where $\delta_x^0\delta_x^0$ has been replaced by a three point δ_{xx}^2. The diagonal dominance of the left hand side of (3.2.7) is guaranteed for any time step and it can be shown that suitable choice of the free parameter ε guarantees stability.

Now, R^* can be identified as a Law-Wendroff operator. However, the Lax-Wendroff scheme has a residual steady state numerical viscosity dependent on the time step size [3.44] and does not seem appropriate for an algorithm designed for use with large time steps. So, R^* is implemented by the following two steps

$$R = - \Delta t \left(A\delta_x^0 u - \nu\delta_{xx}^2 u\right)^n$$

$$\bar{u} = u^n + R \qquad (3.2.8)$$

$$R^* = - \Delta t \left(A\delta_x^0\bar{u} - \nu\delta_{xx}^2\bar{u}\right)$$

This is equivalant to the form of R^* in (3.2.3) in the case of linear coefficients. The two step formation of R^* is analogous to the Brailovskaya scheme [3.44] and has the key advantage that in a steady solution both R and R^* are zero, i.e., the steady solution is independent of the time step size.

We extend the algorithm to the coupled set of three-dimensional Navier-Stokes equations (3.2.1) as follows

$$\left[I - \varepsilon_I \Delta t^2 \lambda_I \delta^2_{II}\right] \left[I - \varepsilon_J \Delta t^2 \lambda_J \delta^2_{JJ}\right] \left[I - \varepsilon_K \Delta t^2 \lambda_K \delta^2_{KK}\right] \Delta \bar{u} = R^* \qquad (3.2.9)$$

where $\bar{u} = (\rho, \rho u, \rho v, \rho w, \rho e)$, $\varepsilon_I, \varepsilon_J$ and ε_K are free parameters (or order unity), λ_I λ_J and λ_K are the spectral radii of the Jacobians associated with the convective fluxes in the I,J and K-wise directions and $\delta^2_{II}\phi$, for example, respresents $(\phi_{I+1} - 2\phi_I - \phi_{I-1})$. The use of spectral radii rather than the Jacobians themselves does not disrupt the formal accuracy or stability of the scheme but allows a significant computer saving since a coupled 5 × 5 system is replaced by five sets of scalar equations. The left hand side is factored into three tridiagonal matrices to allow efficient inversion.

R^* is formed in two steps using the basic finite volume residual

$$R = \sum_{CELL\ FACES} \bar{H} \cdot \overline{dAREA}$$

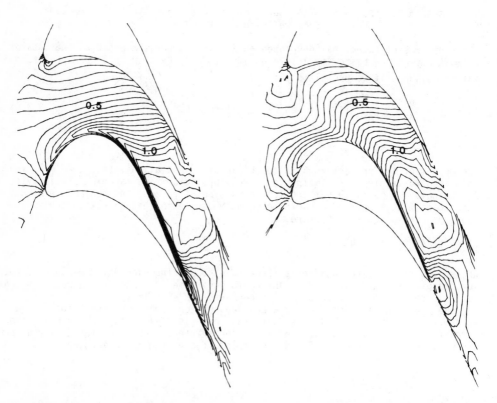

a)   Mach number contours
     at mid span

b)   Mach number contours
     near the endwall

FIGURE 3.2.5 Predicted flow in a linear cascade of gas turbine rotor blades

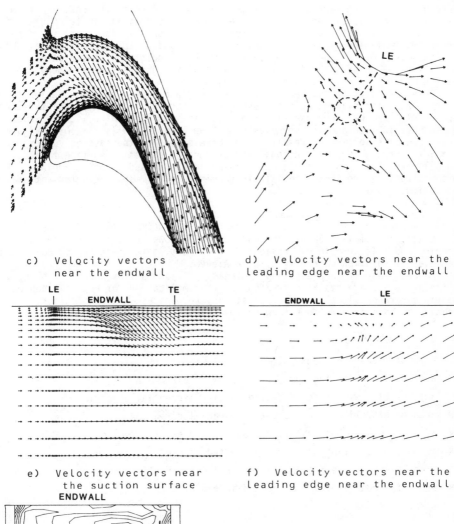

c)   Velocity vectors
     near the endwall

d)   Velocity vectors near the
leading edge near the endwall

e)   Velocity vectors near
     the suction surface

f)   Velocity vectors near the
leading edge near the endwall

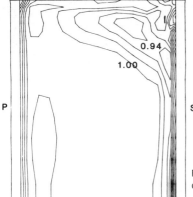

g)   Contours of p_0/p_{01} just
upstream of the trailing edge

FIGURE 3.2.5 Predicted flows in a linear
cascade of gas turbine rotor blades (cont'd)

where $\bar{H}$ contains the various convective fluxes and viscous stresses. To form R the flow field is divided into a large number of six-sided finite volumes with the flow variables stored at cell centers. Fluxes on cell faces are found by linear interpolation of density, velocity, etc., and so the formal spatial accuracy is second order (on smoothly varying meshes) and global conservation ensured. An adaptive second-fourth difference artificial viscosity term recommended by Jameson [3.45] is included in the residual R.

Sample results are presented for the flow in a linear cascade of high turning gas turbine rotor blades; half the symmetrical cascade is modelled and the experimental inlet end wall boundary layer profile is specified. The isentropic exit Mach number is 1.20 and the Reynolds number on axial chord is 8×10^5. Figure 3.2.5 shows a selection of results. The development of the passage vortex and associated strong three dimensional effects near the suction surface towards the end wall are clear. Near the blade leading edge, the saddle point and horseshoe vortex structure are indicated, albeit requiring more mesh points for satisfactory resolution.

3.2.5   Concluding Remarks

The ability to simulate 3D viscous compressible flows in turbomachinery is emerging and this article has reviewed the current state-of-the-art, indicating the level of solution currently being achieved.

Future prospects are exciting. Already 3D viscous flow codes are being used to investigate blade stacking strategies, the implications of end wall contouring and tip clearance flows in compressors; many key issues in turbomachinery are now becoming accessible. We look forward to enhanced engine performance.

NOMENCLATURE

A	area of face of element
ABX	area of bladewise face projected in axial direction
ABT	area of bladewise face projected in circumferential direction
ABR	area of bladewise face projected in radial direction
AQX	area of quasi-orthogonal face projected in axial direction
AQR	area of quasi-orthogonal face projected in radial direction
ASX	area of streamwise face projected in axial direction
ASR	area of streamwise face projected in radial direction
a	speed of sound
C	velocity
CF	correction factor on pressure
Cp	isobaric specific heat capacity
E	internal energy
f	body force
h_t	stagnation enthalpy
$\Delta\ell$	distance
M	Mach number
m	meridional distance
P	static pressure
P_{EFF}	effective static pressure

R   gas constant

RF   relaxation factor

r   radial distance

s   streamwise direction or distance

Δt time step

ΔV control volume

u,v,w   velocity components

x   axial distance

$\delta^{\star}$ boundary layer displacement thickness

θ circumferential angle

γ ratio of specific heats

ν viscosity

ρ fluid density

$\overset{=}{\tau}$ stress tensor

Ω rotation

Subscripts

is   isentropic

m   meridional

r   radial

w   relative

x   axial

θ circumferential

2   downstream

REFERENCES

3.1   Novak, R.A. & Hearsay, R.M.: A Nearly Three Dimensional Intrablade Computing System for Turbomachinery. Parts I & II. *ASME Paper* 76-FE-19; 76-FE-20, 1976.

3.2   Katsanis, T.: Use of Arbitrary Quasi-Orthogonals for Calculating Flow Distribution in a Turbomachine. *ASME Trans., Series A - Journal of Engineering for Power,* Vol. 88, No. 2, April 1966, pp 197-202.

3.3   Denton, J.D.: A Time Marching Method for Two and Three Dimensional Blade to Blade Flow. *ARC* R&M 3775, 1975.

3.4   Bosman, C. & Highton, J.: A Calculation Procedure for 3D Time-Dependent Inviscid, Compressible Flow Through Turbomachinery Blades of any Geometry. *Journal of Mechanical Engineering Science,* Vol. 21, No. 1, 1979.

3.5   Thompkins, W.T.: An Experimental and Computational Study of Flow in a Transonic Compressor Rotor. *Ph.D. Thesis,* M.I.T., June 1976.

3.6   Laskaris, T.E.: Finite Element Analysis of 3D Potential Flow in Turbomachines. *AIAA Journal,* Vol. 16, No. 7, July 1978, pp 717-722.

3.7   Hirsch, C. & Lacor, C.: Rotational Flow Calculations in 3D Blade Passages. *ASME Paper* 82-GT-316, 1982.

3.8   Soulis, J.V.: Calculation of Transonic Potential Flow Through Turbomachinery Blade Rows. *Ph.D. Thesis,* Cambridge University, September 1981.

3.9   Enselme, M.; Brochet, J.; Boisseau, J.P.: Low Cost 3D Flow Computations Using a Mini-system. *AIAA Journal,* Vol. 20, No. 11, November 1982, pp 1515-1520.

3.10   Thompkins, W.T.: A Fortran Program for Calculating 3D, Inviscid, Rotational Flows with Shock Waves in Axial Compressor Blade Rows. *MIT* GT & PDL Report, No. 162, September 1981.

3.11   Denton, J.D.: Extension of the Finite Area Time Marching Method to Three Dimensions. In: *"Transonic Flows in Axial Turbomachinery",* VKI LS 84, February 2-6, 1976.

3.12 Denton, J.D. & Singh, U.K.: Time Marching Methods for Turbomachinery Flow Calculation. In: *"Application of Numerical Methods to Flow Calculations in Turbomachines"*, VKI LS 1979-7, April 23-27, 1979.

3.13 Denton, J.D.: An Improved Time Marching Method for Turbomachinery Flow Calculation. *ASME Paper* 82-GT-239, 1982.

3.14 CAMUS, J-J.; Denton, J.D.; Soulis, J.V.; Scrivener, C.J.: An Experimental and Computational Study of Transonic 3D Flow in a Turbine Cascade. *ASME Paper* 83-GT-12, 1983.

3.15 Barber, T.J.: Analysis of Shearing Internal Flows. *AIAA Paper* 81-0005, 1981.

3.16 Gotthardt, H. & Stark, U.: Theoretical and Experimental Investigations at Conical Turbine Cascades. In: *"Numerical Methods for Flows in Turbomachinery Bladings, Volume 3 : Workshop on 2D and 3D Flow Calculations in Turbine Bladings"*, VKI LS 1982-05, April 26-30, 1982.

3.17 Dawes, W.N. & Squire, L.C.: A Study of Shock Waves in Three Dimensional Transonic Flow. *ASME Trans., Series I - Journal of Fluids Engineering*, Vol. 104, No. 3, September 1982, pp 393-399.

3.18 Freeman, C. & Stow, P.: The Application of Computational Fluid Mechanics to Aero Gas Turbine Compressor Design and Development. *Inst. Mech. Engnrs.*, C70/84, 1984.

3.19 McNally, W.D. & Sockol, P.M.: Review - Computational Methods for Internal Flows with Emphasis on Turbomachinery. *ASME Trans., Series I - Journal of Fluids Engineering*, Vol. 107, No. 1, March 1985, pp 6-22.

3.20 3D Computation Techniques Applied to Internal Flows in Propulsion Systems. *AGARD LS 140, 1985.*

3.21 Peyret, R. & Viviand, H.: Computation of Viscous Compressible Flows Based on the Navier-Stokes Equations. *AGARDograph* 212, 1975.

3.22 Proceedings of the 1980-81 AFOSR/Stanford Conference on Complex Turbulent Flows. Editor P.Kline et al., *Stanford University*, Vols. I, II, III, 1982.

3.23 Patankar, S.V. & Spalding, D.B.: A Calculation Procedure for Heat, Mass and Momentum Transfer in Three-Dimensional Parabolic Flows. *International Journal of Heat & Mass Transfer*, Vol. 15, No. 10, October 1972, pp 1787-1806.

3.24 Briley, W.R.: Numerical Method for Predicting Three Dimensional Steady Viscous Flow in Ducts. *Journal of Computational Physics*, Vol. 14, No. 1, January 1974, pp 8-28.

3.25 Roberts, D.W. & Forester, C.K.: Parabolic Procedure for Flows in Ducts with Arbitrary Cross Sections. *AIAA Journal*, Vol. 17, No. 1, January 1979, pp 33-40.

3.26 Briley, W.R. & McDonald, H.: Analysis and Computation of Viscous Primary and Secondary Flows. *AIAA Paper* 79-1453, 1979.

3.27 Anderson, O.L.: Calculation of Internal Viscous Flows in Axisymmetric Ducts at Moderate to High Reynolds Numbers. *Computers & Fluids*, Vol. 8, No. 4, December 1980, pp 391-411.

3.28 Moore, J. & Moore, J.G.: Calculation of Three Dimensional Flow and Wake Development in a Centrifugal Impeller. *ASME Trans., Series A - Journal of Engineering for Power*, Vol. 103, No. 2, April 1981, pp 367-372.

3.29 Moore, J. & Moore, J.G.: Three Dimensional Viscous Flow Calculations for Assessing the Thermodynamic Performance of Centrifugal Compressors - Study of the Eckardt Compressor. In: "Centrifugal Compressors, Flow Phenomena and Performance". *AGARD CP* 282, May 1980, paper 9.

3.30 Rhie, C.M.; Delaney, R.A.; McKain, T.F.: Three Dimensional Viscous Flow Analysis for Centrifugal Impellers. *AIAA Paper* 84-1296, 1984.

3.31 Khalil, I.M. & Weber, H.G.: Modeling of Three Dimensional Flow in Turning Channels. *ASME Trans., Series A - Journal of Engineering for Gas Turbines & Power*, Vol. 106, No. 3, July 1984, pp 682-691.

3.32 Caretto, L.S.; Gosman, A.D.; Patankar, S.V.; Spalding, D.B.: Two Calculation Procedures for Steady, Three Dimensional Flows with Recirculation. Proc. Third Int. Conf. on Numerical Methods in Fluid Mechanics, Paris, 1972, *Springer-Verlag*, Lecture Notes in Physics, Vol. 19, 1973, pp 60-68.

3.33 Hah, C.: A Navier-Stokes Analysis of Three-Dimensional Turbulent Flows Inside Turbine Blade Rows at Design and Off-Design Conditions. *ASME Trans., Series A - Journal of Engineering for Gas Turbines & Power*, Vol. 106, No. 2, April 1984, pp 421-429.

3.34 Hah, C.: A Numerical Study of Three-Dimensional Flow Separation and Wake Development in an Axial Compressor Rotor. *ASME Paper* 84-GT-34, 1984.

3.35 Hah, C.: A Numerical Modelling of End Wall and Tip-Clearance Flow of an Isolated Compressor Rotor. *ASME Paper* 85-GT-116, 1985.

3.36 Langston, L.S.; Nice, M.L.; Hooper, R.G.: Three Dimensional Flow within a Turbine Cascade Passage. *ASME Trans., Series A - Journal of Engineering for Power*, Vol. 99, No. 1, January 1977, pp 21-28.

3.37 Moore, J. & Moore, J.G.: Performance Evaluation of Linear Turbine Cascades Using Three-Dimensional Viscous Flow Calculations. *ASME Paper* 85-GT-65, 1985.

3.38 Shang, J.G.: An Assessment of Numerical Solutions of the Compressible Navier-Stokes Equations. *Journal of Aircraft*, Vol. 22, No. 5, May 1985, pp 353-370.

3.39 Purohit, S.C.; Shang, J.S.; Hankey, W.L.: Numerical Simulation of Flow Around a Three-Dimensional Turret. *AIAA Journal*, Vol. 21, No. 11, November 1983, pp 1533-1540.

3.40 Warming, R.F. & Beam, R.M.: On the Construction and Application of Implicit Factored Schemes for Conservation Laws. *SIAM-AMS Proceedings*, Vol. 11, 1978.

3.41 Weinberg, B.C. & McDonald, H.: Solution of Three-Dimensional Time Dependent Viscous Flows. Proc. Eighth Int. Conf. on Numerical Methods in Fluid Dynamics; Aachen 1982 *Springer-Verlag*, Lecture Notes in Physics, Vol. 170, 1982.

3.42   Weinberg, B.C.; Yang, R-J.; McDonald, H.; Shamroth, S.J.: Calculation of Two and Three
       Dimensional Transonic Cascade Flow Fields Using the Navier-Stokes Equations. *ASME Paper*
       85-GT-66, 1985.
3.43   Dawes, W.N.: A Pre-Processed Implicit Algorithm for 3D Viscous Compressible Flow.
       6th GAMM Conference on Numerical Methods in Fluid Mechanics, Göttingen, 1985.
3.44   Roache, P.J.: Computational Fluid Dynamics. Albuquerque, *Hermosa Publishers*, 1971.
3.45   Jameson, A. & Baker, T.J.: Multigrid Solution of the Euler Equations for Aircraft
       Configurations. *AIAA Paper* 84-0093, 1984.

CHAPTER 4

Calculation of Wet Steam Stages

4.1 NON-EQUILIBRIUM WET STEAM FLOW IN LOW PRESSURE TURBINES

J.B. Young

4.1.1 Introduction

From the point of view of the designer, wetness effects in steam turbines fall into three broad categories. First and foremost is the need to know the magnitude of departures from equilibrium when turbines operate below the saturation line and whether the resulting changes in pressure and velocity distribution warrant any alteration in blade design. Secondly it is necessary to estimate (and hopefully reduce) the loss in efficiency due to wetness. Thirdly a method is required for predicting the possibility of blade erosion in the machine.

In recent years the emphasis on erosion has declined, the damage to modern LP turbines being less severe than originally predicted. Research interest has mainly centered on the first two categories and some significant advances have been made in non-equilibrium flow calculation methods. Developments in wetness loss prediction have been less successful, however, and it is significant that Baumann's famous 'one per cent for one per cent' rule, first proposed in 1921, is still in common use today.

With the exception of the pioneering work of Gyarmathy [4.1] and Kirillov & Yablonik [4.2] on the structure of two-phase flow in turbines, most of the theoretical work until the mid 1970's was concerned with the prediction of one dimensional nucleating and condensing flows. By and large, excellent agreement between theory and experiment was obtained and it is now generally accepted that the nucleation and droplet growth theories in current use are a satisfactory representation of reality. An excellent review of progress to 1976 can be found in [4.3].

Interesting and necessary as this work was, however, the flow pattern in a low pressure turbine stage with high casing flare and twisted blading is strongly three dimensional and a one dimensional approach is obviously an oversimplification of a very complex geometry. In order to provide useful design tools it was clear that wet steam theory required adaptation for use with conventional turbomachinery calculation methods. The transition from one to two dimensions, however, is by no means straightforward. For example, a typical throughflow calculation may involve the flow in a large number of blade passages in a multi-stage machine and numerical integration procedures for computing the effects of wetness must be implemented a formidable number of times during the course of a calculation. Even with the capabilities of present day computers, this requires the acceleration of conventional techniques by one or even two orders of magnitude.

In the following sections are described some theoretical advances in non-equilibrium flow calculations with which the author has been associated in recent years. After a brief review of the fundamentals of wet steam theory (stressing a unified approach directed towards modern computational methods), the problem of predicting the critical conditions and choking mass flow rate of a transonic wet steam flow is discussed. This is an area of obvious interest to the LP turbine designer and involves the thorny topic of the speed of sound in wet steam. Following this, is a description of a non-equilibrium streamline curvature technique for analysing the flow in LP turbines, together with a number of examples of its application.

4.1.2 Basic Fluid Mechanics of Wet Steam Flows

For an exhaustive discussion of the derivation of the basic equations of wet steam theory, the reader is referred to [4.3]. What follows is a resumé of the main ideas.

4.1.2.1 Gas Dynamic Equations

Wet steam is assumed to be a homogeneous mixture of vapour at pressure p and temperature T_g, and spherical water droplets. The continuous distribution of droplet sizes is discretized into a number of groups such that the ith group contains n_i droplets of mass m_i per unit mass of mixture. The wetness fraction y is then given by

$$y = \Sigma y_i = \Sigma n_i m_i, \qquad\qquad (4.1.1)$$

where the summation is over all the groups. If the vapour density is ρ_g, the mixture density ρ (neglecting the volume of the liquid phase) is given by

$$\rho = \frac{\rho_g}{1-y}. \qquad\qquad (4.1.2)$$

For an established wet steam flow in the absence of nucleation, the distribution of droplet sizes tends to be narrow. Under these conditions, the spectrum may be represented by a single droplet group of average radius.

Departures from equilibrium manifest themselves in two distinct ways. *Thermal non-equilibrium* is characterized by a temperature difference between droplets and vapour and also by the fact that the wetness fraction does not correspond to the equilibrium value at the prevailing pressure. *Inertial non-equilibrium* gives rise to velocity differences between the phases. Both phenomena have associated relaxation times which depend, among other things, on the droplet radius r, as shown in figures 4.1.1 and 4.1.2. For the range of fog droplet sizes found in steam turbines (0.05 μm < r < 1.0 μm) the inertial relaxation time is at least one order of magnitude less than the thermal relaxation time. It is usually acceptable, therefore, to neglect the velocity slip between the phases and assume the flow to remain in inertial equilibrium.

Making this approximation, the continuity and momentum equations for steady, inviscid flow can be written

$$\nabla \cdot (\rho \vec{C}) = 0, \qquad\qquad (4.1.3)$$

and

$$(\vec{C} \cdot \nabla)\vec{C} + \frac{\nabla p}{\rho} = 0, \qquad\qquad (4.1.4)$$

where $\vec{C}$ is the common velocity of the phases. In the absence of external heat and work transfers, the energy equation also takes the familiar form,

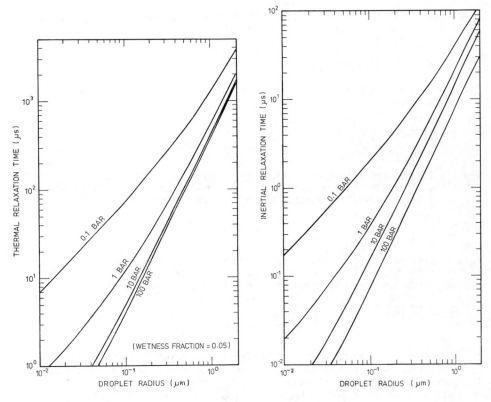

FIGURE 4.1.1 Thermal relaxation time of monodispersed water droplets in steam

FIGURE 4.1.2 Inertial relaxation time of monodispersed water droplets in steam

$$\nabla \cdot \left[\rho \vec{C} \left(h + \frac{C^2}{2} \right) \right] = 0, \tag{4.1.5}$$

where h is the mixture specific enthalpy,

$$h = (1-y)h_g + \Sigma y_i h_i. \tag{4.1.6}$$

h_g and h_i are the specific enthalpies of the vapour and liquid phases respectively and must be evaluated at the corresponding temperatures T_g and T_i.

Combination of equations (4.1.3) and (4.1.5) show that the stagnation specific enthalpy

$$h_t = (1-y)h_g + \Sigma y_i h_i + \frac{C^2}{2}, \tag{4.1.7}$$

is constant along a streamline. For turbomachinery calculations this can be extended to the definition of the rothalpy,

$$I = h_t - r_s \Omega C_\theta, \tag{4.1.8}$$

which is constant along a stream surface in a rotating blade passage.

In (4.1.8) r_s is the radial position of the streamline measured from the turbine axis, Ω is the rotational speed and C_θ is the tangential component of the absolute steam velocity.

4.1.2.2 Steam Properties

The equation of state of the vapour phase can be expressed quite generally in differential form by

$$\frac{d\rho_g}{\rho_g} = (\beta p)\,\frac{dp}{p} - (\alpha T_g)\,\frac{dT_g}{T_g}, \tag{4.1.9}$$

where α is the coefficient of thermal expansion and β is the isothermal compressibility. Although it is perfectly possible to develop the theory in terms of α and β, it is convenient to assume the *vapour phase* obeys the perfect gas law, thus implying that

$$\alpha T_g = \beta p = 1. \tag{4.1.10}$$

This simplification aids understanding and is in any case an acceptable representation of the properties of steam at the low pressures to be found in the final stages of the turbine. All numerical results presented below, however, have been obtained using an accurate equation of virial form, full details of which can be found in [4.1.4].

4.1.2.3 Nucleation

In order to calculate the wetness fraction at any point in the flow field, it is necessary to specify the number of droplets present. When steam condenses in laboratory nozzles, it is known that droplets are formed by spontaneous nucleation in the core flow. In turbines, however, it is possible that high turbulence levels and condensation in blade wakes may affect the process, but no experimental data is available to confirm or refute these hypotheses. Normally, therefore, it is assumed that the vast majority of droplets are formed by spontaneous nucleation in the main bulk of the flow.

A lucid description of classical nucleation theory can be found in [4.5] where it is shown that the rate of formation of nuclei per unit mass of steam is given by

$$J_{CL} = q_c \sqrt{\frac{2\sigma N_0^3}{\pi M^3}}\,\frac{\rho_g}{\rho_\ell}\,\exp\left[-\frac{4\pi r_\star^2 \sigma}{3kT_g}\right], \tag{4.1.11}$$

where the Kelvin-Helmholtz critical radius $r_\star$ is calculated from

$$r_\star = \frac{2\sigma T_s}{\rho_\ell h_{fg}\Delta T}. \tag{4.1.12}$$

All symbols are defined in the nomenclature, but particular attention should be paid to the supercooling

$$\Delta T = T_s - T_g, \tag{4.1.13}$$

which is the 'driving potential' for nucleation and droplet growth.

Laboratory nozzle tests [4.4] indicate that equation (4.1.11) is in error for steam at low pressures. A more accurate version is obtained by applying a correction due to Kantrowitz [4.6] for non-isothermal effects. The recommended expression for the nucleation rate J is thus given by

$$J = \frac{1}{(1+\psi)} J_{CL},$$ (4.1.14)

where

$$\psi = 0.26 \left(\frac{h_{fg}}{RT_g}\right)\left(\frac{h_{fg}}{RT_g} - 0.5.\right).$$ (4.1.15)

4.1.2.4 *Droplet Growth*

Condensation on a liquid droplet proceeds at a rate governed by the ability of the vapour to conduct the latent heat away from the droplet surface. The generally accepted form of the droplet growth equation is that due to Gyarmathy [4.3].

$$h_{fg} \frac{dr_i}{dt} = \frac{\lambda_g}{r_i \rho_i (1+3.78\, Kn_i/Pr_g)} (T_i - T_g),$$ (4.1.16)

where r_i, T_i, ρ_i and Kn_i are the radius, temperature, density and Knudsen number of droplets of the ith group, λ_g and Pr_g are the thermal conductivity and Prandtl number of the vapour and d/dt represents the time derivative following a particular droplet. Equation (4.1.16) appears to be accurate over most of the range of interest, although it is possible that modifications are required at low pressure [4.4].

By combining a mass transfer equation with (4.1.6), Gyarmathy also showed that the droplet temperature is related to the vapour supercooling by

$$T_i - T_g = \Delta T - \Delta T_i,$$ (4.1.17)

where ΔT_i is the *capillary supercooling*,

$$\Delta T_i = \frac{2\sigma T_s}{\rho_i h_{fg} r_i}.$$ (4.1.18)

Except for very small, freshly nucleated, droplets, the capillary supercooling may be neglected ($\Delta T_i \simeq 1K$ for $r = 0.02\ \mu m$).

4.1.2.5 *Calculation of the Wetness Fraction*

The contribution to the wetness fraction from the ith droplet group is

$$y_i = n_i m_i = \frac{4}{3} \pi r_i^3 \rho_i n_i.$$ (4.1.19)

In a non-nucleating flow the n_i remain constant. Combination of equations (4.1.16), (4.1.17) and (4.1.19) then gives

$$\frac{dy_i}{dt} = \frac{(1-y)c_{pg}}{h_{fg}} \frac{(T-T_i)}{\tau_i}, \tag{4.1.20}$$

where τ_i is the *thermal relaxation time* defined by

$$\tau_i = \frac{(1-y)c_{pg}r_i^2\rho_i(1+3.78\ Kn_i/Pr_g)}{3y_i\lambda_g}. \tag{4.1.21}$$

Defining an average relaxation time τ by

$$\frac{1}{\tau} = \Sigma\ \frac{1}{\tau_i}, \tag{4.1.22}$$

and an average capillary supercooling ΔT_ℓ by

$$\frac{\Delta T_\ell}{\tau} = \Sigma\ \frac{\Delta T_i}{\tau_i}, \tag{4.1.23}$$

it follows that

$$\frac{dy}{dt} = \Sigma\ \frac{dy_i}{dt} = \frac{(1-y)c_{pg}(\Delta T - \Delta T_\ell)}{h_{fg}\ \tau} \tag{4.1.24}$$

The wetness fraction can now be obtained by integrating (4.1.24) along a streamline.

   Equations (4.1.22) and (4.1.23) demonstrate how a non-nucleating polydispersed flow can be represented by an equivalent monodispersion. The technique can be extended to nucleating flows as the contribution to dy/dt from the freshly nucleated droplets is always negligible. However, the relaxation time τ changes very rapidly in a nucleating flow and the integration must be carried out using very small time increments.

   4.1.2.6   The Supercooling Equation

The main objective of any non-equilibrium calculation procedure is to predict the distributions of the supercooling and wetness fraction. By combining equations (4.1.3)-(4.1.7) with (4.1.24) (for details see [4.7]), it is possible to derive an explicit equation for the variation of supercooling along a streamline,

$$\frac{d}{dt}(\Delta T) + \frac{\Delta T}{\tau} = F\ \frac{1}{p}\ \frac{dp}{dt}, \tag{4.1.25}$$

where

$$F = \frac{RT_s}{(1-y)c_{pg}} \left[\frac{cT_s}{h_{fg}} - (1-y)\ \frac{T_g}{T_s}\right]. \tag{4.1.26}$$

Because F is a slowly varying function, (4.1.25) can be integrated analytically over regions where the rate of expansion (1/p)(dp/dt) and the relaxation time τ remain constant. The result is

$$\Delta T = \Delta T_0 \; e^{-t/\tau} + F\tau \left(\frac{1}{p}\frac{dp}{dt}\right)\left(1-e^{-t/\tau}\right).$$
(4.1.27)

where ΔT_0 is the supercooling at $t = 0$.

Figures 4.1.3 and 4.1.4 show the variation of the supercooling with pressure for various droplet sizes and expansion rates, calculated from (4.1.27). As τ is assumed to remain constant when integrating (4.1.22), the results only apply to non-nucleating wet steam flows with approximately constant expansion rates. Despite these restrictions, however, the calculations give good insight into the underlying physical processes by demonstrating the whole range of behaviour from equilibrium to frozen flow. This is pursued at some length in [4.7].

Even when the expansion rate varies, (4.1.27) can be applied stepwise through the flow field to calculate the variation of supercooling. Its great advantage lies in the fact that the mathematical stiffness inherent in equation (4.1.25) through the (possibly) small values of τ has been removed, thus allowing much larger integration steps to be taken while still retaining numerical stability. Many of the difficulties of early one dimensional calculation procedures for wet steam flows (which always involved high CPU usage) stemmed from problems involving stiffness and this is a very convenient method of overcoming them.

4.1.2.7 *Thermodynamic Wetness Loss*

The interphase temperature difference causes an irreversible heat and mass transfer resulting in an overall entropy increase referred to as the *thermodynamic wetness loss*. This is partially responsible for the reduction in work output of turbines operating in the two-phase flow region.

The derivation of an expression for the rate of entropy increase along a streamline can be found in [4.4]. The result is

$$\frac{ds}{dt} \approx h_{fg}\frac{\Delta T}{T_s^2}\frac{dy}{dt},$$
(4.1.28)

where s is the mixture specific entropy.

Equations (4.1.24) and (4.1.28) can be integrated analytically over regions where the relaxation time and expansion rate remain constant by substituting equation (4.1.27) for ΔT. It is thus possible to calculate the total thermodynamic entropy increase Δs for the nozzle or blade passage. The *thermal energy loss coefficient* is then defined by

$$\xi_T = \frac{T_s \Delta s}{(C_e^2/2)}$$
(4.1.29)

where C_e is the flow exit velocity. The variation of ξ_T with nozzle pressure ratio is shown in figures 4.1.5 and 4.1.6 for the special case of constant relaxation time and rate of expansion. The results give a good quantitative indication of the effect of droplet size.

Despite the emphasis laid on the thermodynamic effect, it must not be forgotten that there are other important components of the wetness loss. These include droplet drag, deposition and centrifuging losses, but apart from some early work [4.1,4.2], little attempt has been made to estimate their magnitude.

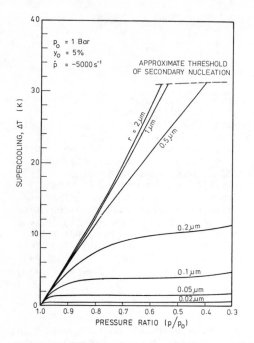

FIGURE 4.1.3 Development of supercooling during an expansion. (Various droplet sizes, one expansion rate)

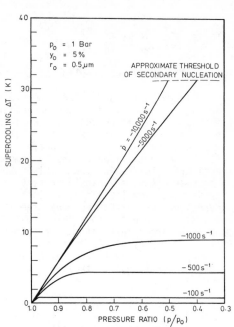

FIGURE 4.1.4 Development of super-cooling during an expansion. (Various expansion rates, one droplet size)

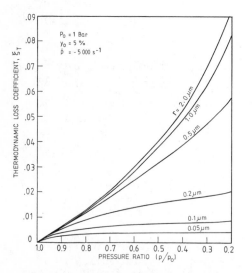

FIGURE 4.1.5 Variation of the thermodynamic loss coefficient with nozzle pressure ratio. (Various droplet sizes, one expansion rate)

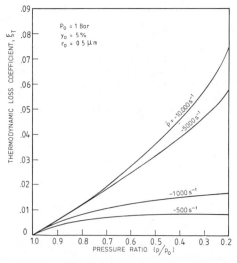

FIGURE 4.1.6 Variation of the thermodynamic loss coefficient with nozzle pressure ratio. (Various expansion rates, one droplet size)

4.1.3 Choking in Non-Equilibrium Flows

4.1.3.1 *The Mass Flow Rate in Unchoked Flows*

The early experiments of Stodola [4.8] showed that an important effect of super-saturation in wet steam flows is to increase the swallowing capacity of nozzles by up to 10% compared with the value calculated from equilibrium theory. The effect is measured in terms of the *mass flow coefficient* ϕ, defined by

$$\phi = \frac{\text{actual mass flow rate}}{\text{equilibrium mass flow rate}} = \frac{\dot{m}}{\dot{m}_e}, \tag{4.1.30}$$

where $\dot{m}_e$ is the mass flow rate calculated assuming complete thermal and inertial equilibrium during the expansion. The phenomenon is important for turbine design because the power output of the machine is proportional to the throughput.

The mass flow coefficient is theoretically a function of droplet size and this has recently been confirmed in a series of tests conducted by Filippov et al. [4.9]. In these experiments, a cascade of turbine blades operating at a constant pressure ratio was supplied with both wet and dry steam of varying conditions, the mass flow rate being measured in each case. The results of two series of tests with nominal cascade exit Mach numbers of 0.65 and 0.9 are reproduced in figure 4.1.7.

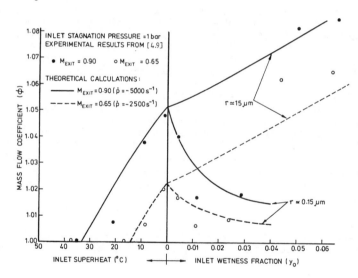

FIGURE 4.1.7 Theoretical and experimental determination of the mass flow coefficient in an unchoked turbine cascade

The wet steam used in the tests was supplied from two different sources. In one case water was injected into the flow through a number of hypodermic needles giving an average droplet radius of 15 μm. The mass flow coefficient was then observed to rise with increasing wetness fraction. In the second case the cascade was fed with wet steam nucleated in a turbine stage located further upstream and the droplet radius, as measured by a light scattering technique, was found to be 0.15 μm. Under these circumstances the mass flow co-efficient fell with increasing wetness fraction.

A theoretical comparison based on the data from these experiments was performed by the present writer and a full description can be found in [4.10].

Assuming one dimensional flow at cascade exit, the mass flow rate is given by

$$m = \rho AC = \frac{\rho_g AC}{(1-y)} ,$$ (4.1.31)

where A is the flow cross-sectional area and all variables are evaluated in the throat plane of the cascade. In order to evaluate the variables in (4.1.31), it was necessary to estimate the supercooling at the throat of the cascade. This was obtained by integrating equation (4.1.25) from the known cascade inlet conditions to the specified exit pressure, the rate of expansion being evaluated from the (admittedly rather meagre) geometrical details to be found in [4.9]. The wetness fraction was then obtained by integrating equation (4.1.24) and was completely defined by the exit pressure and supercooling through the equation of state for the vapour phase. Finally, the flow velocity was calculated from the energy equation (4.1.7).

The results of the calculations are presented in figure 4.1.7, where it can be seen that the overall agreement between theory and experiment is excellent. It is particularly encouraging that wet steam theory so faithfully predicts the very different behaviour exhibited by flows containing large and small droplets.

4.1.3.2 *Critical Conditions in Choked Flows*

Unfortunately the work described in [4.9] did not include mass flow rate measurements when the cascade was choked. In fact the only published measurements of the choking mass flow rate in wet steam are those reported in [4.11] and, as the droplet size was very large, they are only of value for corroborating the theory in the comparatively uninteresting limiting case of frozen flow. The theoretical treatment that follows, therefore, although giving plausible results, is largely unsubstantiated.

Equation (4.1.31) can be used to calculate the mass flow rate in a choked nozzle if all the parameters are evaluated at the critical condition. In a single phase flow this is easily shown to occur at the nozzle throat where the flow velocity equals the local speed of sound, but in two phase flow the speed of sound varies with the liquid dispersion and also with the frequency of the wave itself. According to Deich [4.12], Konorski [4.13] and Petr [4.14], two limiting cases can be identified. Very high frequency waves moving through a medium containing large droplets propagate at the *frozen* speed of sound

$$a_f = (k_f p/\rho_g)^{1/2} ,$$ (4.1.32)

where k_f is the isentropic exponent of the *vapour phase alone*. On the other hand, very low frequency waves moving through a medium containing small droplets propagate at the *equilibrium* speed of sound

$$a_e = (k_e p/\rho)^{1/2} ,$$ (4.1.33)

where k_e is the Zeuner or equilibrium isentropic exponent of the *mixture*.

The question of the definition of the critical velocity in a steady flow now arises and whether this can be related to the frequency dependent speed of sound. Konorski [4.15,4.16] suggested that the critical velocity should be defined as the velocity at the nozzle throat when the flow is choked. Unfortunately, the extension to two dimensional flow is not obvious and a better *definition* of critical velocity is that *vapour phase* velocity C_g which is locally equal to the *frozen speed of sound* af. *Defining* a Mach number by

$$M = \frac{C_g}{a_f} ,$$ (4.1.34)

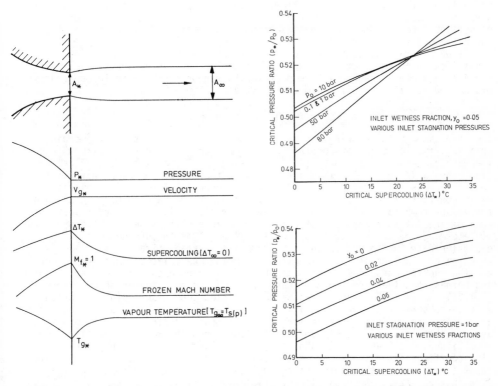

FIGURE 4.1.8 Variation of the flow parameters in a wet steam jet emerging from a choked converging nozzle (schematic diagram only)

FIGURE 4.1.9 Critical pressure ratio as a function of critical supercooling and nozzle inlet conditions

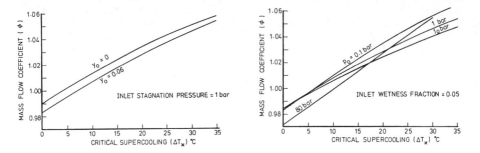

FIGURE 4.1.10 Mass flow coefficient for a converging nozzle

it can be shown that a wet steam flow is choked if, and only if, a Mach number
of unity exists somewhere in the flow. Only under certain conditions, however,
does this occur at the nozzle throat and in other cases the critical condition
occurs downstream in the diverging section. In the latter circumstances the
throat velocity is subcritical.

The mathematical justification of these statements can be found in [4.10]
and follows from the fact that the leading tip of any disturbance travels with
the frozen speed of sound with respect to the vapour phase quite independently
of the degree of non-equilibrium of the steam.

4.1.3.3 *Choking in Converging Nozzles*

The simplest case of choking occurs in a converging nozzle, the jet emerging
into an infinite region of constant static pressure. Under these circumstances,
the Mach number, as defined by (4.1.34), is unity in the exit plane of the nozzle. The variation of the flow parameters in the jet as the steam reverts to
equilibrium is illustrated schematically in figure 4.1.8. The jet velocity remains constant at the critical value $C_{g\star}$, although the Mach number decreases as
the vapour temperature rises towards the saturation value. The cross sectional
area of the jet increases downstream of the nozzle exit plane by a factor which
is determined by the critical supercooling $\Delta T_{\star}$. This in turn depends on the
expansion rate in the nozzle and the thermal relaxation time.

The critical pressure ratio $p_{\star}/p_0$ (p_0 being the inlet stagnation pressure)
and the choking mass flow rate $\dot{m}_{\star}$ can be evaluated once the critical supercooling
and wetness fraction are known. These are obtained by integrating equations
(4.1.24) and (4.1.25) from the known nozzle inlet conditions. The variation in
exit wetness fraction for a specified exit supercooling is very small and results
are, therefore, presented in terms of the exit supercooling only, it being
assumed that this effectively defines the corresponding wetness fraction.

Figures 4.1.9 and 4.1.10 show the critical pressure ratio and choking mass
flow coefficients as functions of the critical supercooling for various nozzle
inlet conditions. It is interesting to note that for $\Delta T_{\star} = 0$ (near-equilibrium
flow in the nozzle) the mass flow coefficient is slightly less than one. This
surprising result follows from the fact that the critical velocity *always* equals
the *frozen* speed of sound, even if the departure from equilibrium during the
expansion is infinitesimally small [4.10].

4.1.3.4 *Choking in Converging-Diverging Nozzles*

Choking in a converging-diverging nozzle is more complex because the swallowing
capacity may be limited by the nozzle geometry downstream of the throat rather
than by the minimum cross section area [4.17]. This is easily appreciated by
a study of figure 4.1.8. Any diverging section added to the nozzle must diverge
at such an angle so as not to restrict the jet in any way. If this is not so,
the critical condition (M = 1) is established in the diverging section and
the throat Mach number is less than unity.

A condition ensuring that the critical velocity occurs at the geometrical
throat is derived in [4.10]. For plane two dimensional nozzles with throat
width $w_{\star}$, the angle of divergence at the throat $\theta_{\star}$ must satisfy

$$\tan\theta_{\star} > \left(1 - \frac{c_{pg}T_g}{h_{fg}}\right)\left(\frac{w_{\star}}{C_{g\star}\tau}\right)\frac{\Delta T_{\star}}{T_{g\star}} \qquad (4.1.35)$$

If this condition is fulfilled, then the calculation of the choking mass flow
rate is identical to that for a converging nozzle with the same throat area.
On the other hand, if the angle of divergence is less than the limiting value
of equation (4.1.35), the throat velocity and choking flow rate are reduced.
In practice the latter condition is more often realized than the former.

Due to the infinite variety of nozzle shapes, no simple analytical theory exists to describe such cases which are best computed numerically. Typical examples can be found in [4.17]. It is possible, however, to analyse the case of a long frictionless parallel duct fed by a converging nozzle, which represents the limiting case of a converging-diverging nozzle with zero angle of divergence. Apart from noting that the possible flow patterns are both complex and intriguing, no details will be presented here. For a full discussion the reader is referred to [4.10].

4.1.4 Throughflow Calculations in LP Turbines

4.1.4.1 Theory

Axisymmetric throughflow calculations are the backbone of turbine design because they give reliable predictions of flow distribution from very limited geometrical details of the blading. The streamline curvature method has emerged as the most popular technique for integrating the flow equations and good accounts of the theory can be found in [4.18 and 4.19], (see also chapter 2.1). However, although complicated models for the loss prediction are nowadays incorporated in the procedures, it is invariably assumed that the steam remains in thermodynamic equilibrium in the two-phase region. Consequently the calculations give no information on droplet size, wetness loss or changes in pressure distribution due to steam supercooling.

In order to overcome these shortcomings, a non-equilibrium throughflow calculation method has been developed [4.20,4.21]. This is based on the approach for single-phase and equilibrium flow described in [4.22] which has the advantages of being able to deal with fully choked multi-stage turbines and of employing a more realistic representation of the blade leading and trailing edges.

As in all throughflow methods the calculations are performed in the meridional plane under the assumption of axial symmetry. It is also assumed that the liquid and vapour phases move at the same velocity and that the droplet spectrum can be represented by an equivalent monodispersion as described previously.

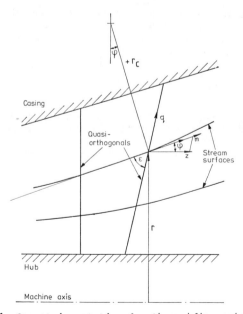

FIGURE 4.1.11   Geometric notation for throughflow calculations

A typical calculation grid is shown in figure 4.1.11 and is composed of straight *quasi-orthogonals* (QO's) running from hub to casing and curved *stream surfaces*. Initially the latter are divided uniformly, but converge on the true stream surfaces as the calculation proceeds.

The essence of the method is the simultaneous solution of the momentum equation in the QO direction, together with the continuity equation and the Euler turbine equation applied along the stream surfaces.

The two-phase, near-radial momentum equation (4.1.20) can be written

$$\frac{\partial}{\partial q}\left(\frac{C_m^2}{2}\right) = \frac{\partial h_t}{\partial q} - T_g\frac{\partial s}{\partial q} - \frac{1}{2r_s^2}\frac{\partial}{\partial q}(r_s C_\theta^2) + \frac{C_m^2}{r_c}\sin(Q-\varphi) + C_m\frac{\partial C_m}{\partial m}\cos(Q-\varphi)$$

$$+ h_{fg}\frac{\Delta T}{T_s}\frac{\partial y}{\partial q}, \tag{4.1.36}$$

where C is the *absolute* velocity, q, m and θ represent the QO, meridional and circumferential directions respectively, h_t is the absolute stagnation enthalpy and s is the mixture entropy defined by

$$s = (1-y)s_g + ys_\ell. \tag{4.1.37}$$

Equation (4.1.36) is identical to the single-phase radial momentum equation with the exception of the final term.

Using a suitable finite difference scheme, (4.1.36) can be integrated numerically along the QO (A-B in Fig. 4.1.11), the constant of integration being specified by the continuity requirement,

$$\dot{m} = \int_A^B 2\pi r_s (1-b)\rho C_m \sin(Q-\varphi) \ dq, \tag{4.1.38}$$

where b is the fractional reduction in annulus area due to blade blockage. ρ is the mixture density defined by (4.1.2).

In duct regions between blade rows the angular momentum of the fluid along each stream surface is conserved,

$$\frac{\partial}{\partial m}(r_s C_\theta) = 0. \tag{4.1.39}$$

For flow within blade rows C_θ is determined from C_m, the relative flow angle β and the blade speed. Values of β at the blade trailing edges and within the passages are specified as input data.

In the streamwise direction the rothalpy $(I = h_t - r_s\Omega C_\theta)$ is constant,

$$\frac{\partial I}{\partial m} = 0. \tag{4.1.40}$$

The entropy increase in the streamwise direction is conveniently divided into two components, the first due to aerodynamic and viscous losses and the second to thermodynamic irreversibilities. Thus

$$\frac{\partial s}{\partial m} = \frac{\partial s_A}{\partial m} + \frac{\partial s_T}{\partial m}. \tag{4.1.41}$$

Aerodynamic entropy increases are calculated from loss coefficients ξ_A which are either specified as input data or computed from actual flow conditions at run time. The thermodynamic entropy increase is given by equation (4.1.28). Strictly speaking, other components of the wetness loss, for example that due to droplet drag, should be included in (4.1.41), but routines for calculating these effects have not yet been developed.

The main difference between an equilibrium and non-equilibrium streamline curvature procedure is in the calculation of the supercooling and wetness fraction. If the stream is at equilibrium the supercooling is always zero and the wetness fraction is completely defined by the local pressure and entropy. In non-equilibrium flow, however, ΔT and y must be calculated by integrating the nucleation and droplet growth equations (4.1.14, 4.1.16, 4.1.17) along the stream surfaces between quasi-orthogonals.

4.1.4.2 *Method of Solution*

The basis of the numerical scheme for the simultaneous solution of the flow equations is similar to that described in [4.22].

Iteration commences by making an initial guess of the stream surface positions. A three point parabolic curve fit is then used to estimate the slopes and curvatures necessary for integrating equation (4.1.36).

The solution proceeds by marching downstream from one QO to the next. Depending on whether the current QO is in a duct region or blade passage, one of the following sequences of calculations is performed.

(a) <u>Duct flow</u> : Denoting the current QO by subscript J, it follows from (4.1.39), (4.1.40) and (4.1.41)

$$\left. \begin{array}{l} (r_s c_\theta)_J = (r_s c_\theta)_{J-1} \\[2mm] (h_t)_J = (h_t)_{J-1} \\[2mm] s_J = s_{J-1} + \Delta s_T \end{array} \right\} \tag{4.1.42}$$

Aerodynamic losses are assumed to be zero outside the blade row, but thermodynamic losses are included.

(b) <u>Blade flow</u> : Stators and rotors are treated identically, except that the angular speed Ω is set equal to zero for the former. Thus

$$\left. \begin{array}{l} (r_s c_\theta)_J = \left[r_s \left(C_m \sin(Q-\varphi)\tan\beta + r_s\Omega \right) \right]_{J-1} \\[2mm] (h_t)_J = (h_t)_{J-1} + \Omega \left[(r_s c_\theta)_J - (r_s c_\theta)_{J-1} \right] \\[2mm] s_J = s_{J-1} + \Delta s_A + \Delta s_T. \end{array} \right\} \tag{4.1.43}$$

The relative flow angle β is specified data. At blade exit

$$\cos\beta = \frac{\text{opening}}{\text{pitch}} \tag{4.1.44}$$

if the flow is subsonic with respect to the frozen speed of sound. If the velocity is supersonic the deviation from the blade angle must be calculated and Denton's method [4.22] is employed. Essentially this seeks to satisfy the continuity equation between the throat and trailing edge of a blade and requires the specification of a suitable isentropic exponent. Following the discussion of section (4.1.3), it is evident that the isentropic exponent of the vapour phase alone is applicable and accordingly a value of 1.32 was adopted.

In order to calculate the change in wetness fraction between quasi-orthogonals it is necessary to integrate the nucleation and droplet growth equations along the stream surface. Generally, the small thermal relaxation times preclude the possibility of applying the equations in finite difference form between adjacent quasi-orthogonals directly and the stream surface must be split into shorter sections for this part of the calculation.

The integration of equation (4.1.25) requires the specification of the variation of expansion rate. Usually detailed blade geometry is not available and it was therefore assumed that the circumferential component of the relative velocity varies in a parabolic fashion from the leading edge to trailing edge with a condition of zero loading imposed at the leading edge. This appears to be a reasonable assumption, but alternative possibilities should be considered to investigate the sensitivity of the results to the model adopted.

The wet steam integration routines furnish values of supercooling, wetness fraction and thermodynamic entropy increase at the current QO and these, together with equations (4.1.42) and (4.1.43) completely define the pressure, vapour phase temperature and density. With these updated property values, the radial momentum and continuity equations can be solved to give an improved meridional velocity distribution along the QO. The iterative process continues until a specified convergence criterion is satisfied. Damping of changes is necessary to stabilize the numerical scheme, but does not affect the final solution.

Special procedures are required when the turbine is fully choked. Under these circumstances the mass flow rate cannot be found by gradual small adjustments without incurring numerical instability. However, a method has been devised by Denton which can be applied to non-equilibrium calculations in a virtually identical manner. For further details the reader is referred to [4.22].

4.1.4.3   The Primary Nucleating Stage

In a large six stage LP turbine, the steam usually crosses the saturation line in the vicinity of stage four. Because nucleation theory predicts a delay in the precipitation of moisture, this *primary nucleating stage* offers a good starting point for studying the differences between wet equilibrium and dry supercooled operation.

Figure 4.1.12 shows the calculation grid for the fourth stage of a low pressure turbine of a 500 MW set. The casing boundary stream surface has been simplified to avoid numerical problems associated with slope discontinuities. The stream surfaces plotted in figure 4.1.12 are those obtained from the converged non-equilibrium solution, but they differ only slightly from those calculated assuming equilibrium flow. It is worth noting that this statement could apply to any of the turbines analyzed to date.

The objective of the exercise was to estimate the likely magnitude of non-equilibrium effects and, accordingly, the results are presented as a comparison between equilibrium and non-equilibrium analyses. Experimental verification is difficult, but armed with this type of information the designer can assess whether the extra complexity of a non-equilibrium calculation is warranted or not.

The differences between the two calculations appear, among other things, as differences in the stage pressure ratio and mass flow rate and depend on the influence of the adjoining stages. However, in order to demonstrate the salient points the stage was considered in isolation and two, albeit artificial, test cases were computed. In the first, equal mass flow rates for both equilibrium and non-equilibrium solutions were specified, while in the second, stage pressure ratio was specified as constant.

(a) Specified mass flow rate. At the stage inlet the steam was assumed to be at the design pressure of 0.7 bar and superheated by 7 K. Aerodynamic losses were not included.

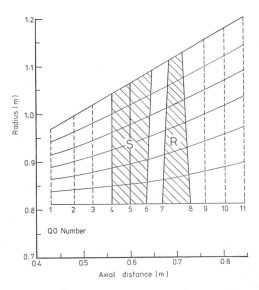

FIGURE 4.1.12   Computational grid for stage 4 of a 500 MW, low pressure turbine

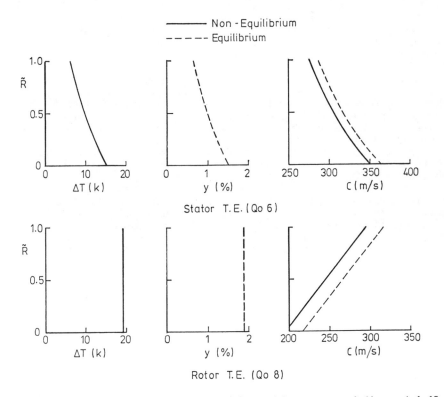

FIGURE 4.1.13   Theoretical results for the turbine stage   of figure 4.1.12
for a prescribed mass flow rate

Figure 4.1.13 shows a comparison between the two solutions at the blade trailing edges. At stage outlet the equilibrium wetness is about 2%, but the non equilibrium calculation suggests that the steam remains dry and supercooled throughout the stage. Supercooled steam is denser than equilibrium steam at the same pressure and gives rise to lower velocities for the same mass flow rate. This results in a reduced flow incidence angle onto both stator and rotor blades as shown in figure 4.1.14. In the present example the deviation from equilibrium incidence is between 5° and 10° depending on the spanwise position.

The reduced relative velocities result in a 14% deficit in power output as compared with the equilibrium calculation. The stage pressure ratio also falls slightly from 1.56 for equilibrium to 1.48 for non-equilibrium flow.

(b) Specified stage pressure ratio. The second test was computed with a stage pressure ratio of 2.13 and aerodynamic loss coefficients of ξ_A = 0.1. The results are shown in figure 4.1.15.

The expansion through the stator row is similar to the previous case except that the equilibrium velocity at the trailing edge is now lower than the non-equilibrium value. This is due to differences in mass flow rate counteracting the effect of increased density due to supercooling and, indeed, the mass flow coefficient as defined by (4.1.30) is calculated to be 1.045.

In expanding through the rotor the steam becomes progressively more super-cooled and high nucleation rates occur at the trailing edge. In the following duct region the droplets grow rapidly to an average radius of 0.02 μm. The wet-ness fraction from the two solutions are then indistinguishable showing that equilibrium has been completely re-established.

The thermodynamic loss coefficient for the rotor was calculated to be ξ_T = 0.07 and the stage power output as calculated by the non-equilibrium analy-sis was 6% higher than the equilibrium value. This should be contrasted with the reduction of 14% found in the first test calculation and demonstrates that non-equilibrium effects can be responsible for comparatively large variations in performance.

4.1.4.4 *The Effect of Droplet Size on Last Stage Performance*

Calculations have also been performed to investigate the effect of droplet size on the performance of the final stage of a typical LP turbine [4.23]. The basis for the calculations was a one-third scale model of stage 6 of the same 500 MW turbine and the geometry is shown in figure 4.1.16.

Three cases corresponding to monodispersed droplets of radii 0.02, 0.1, and 0.2 μm, uniformly distributed at stage inlet, were computed. The inlet wetness was 5.4% and the steam was assumed to be at equilibrium at this point. A stage pres-sure ratio of 3.56 was specified and suitable aerodynamic loss coefficients were also prescribed. The results of the calculations for droplets of radii 0.02 μm and 0.1 μm are shown in figure 4.1.17, together with the equilibrium solution.

In expanding through the stator row the steam becomes progressively more supercooled and this is particularly pronounced at the blade root where the ex-pansion rate is high. Near the hub the steam carrying the droplets of radius 0.1 μm attains a supercooling of 34 K at the trailing edge and this is suffi-ciently high to trigger a secondary nucleation.

Secondary nucleations are awkward to deal with computationally. Ideally two average droplet radii should be retained in the calculations to represent the primary and secondary groups and the growth rates calculated separately. In the work being described, however, it was more convenient to average the freshly nucleated droplets with those already existing and retain one group only. This explains the dramatic decrease in average droplet radius near the hub shown in figure 4.1.17.

By the leading edge of the rotor (Q06) equilibrium has been re-established over most of the annulus, but the steam again becomes supercooled during its passage through the moving blade. Note that the deviation from equilibrium in-creases with droplet size due to a reduction in liquid surface area available

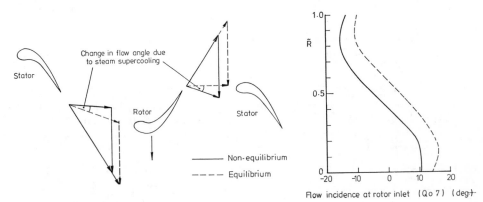

FIGURE 4.1.14 Effect of supercooling on flow incidence for a given mass flow rate

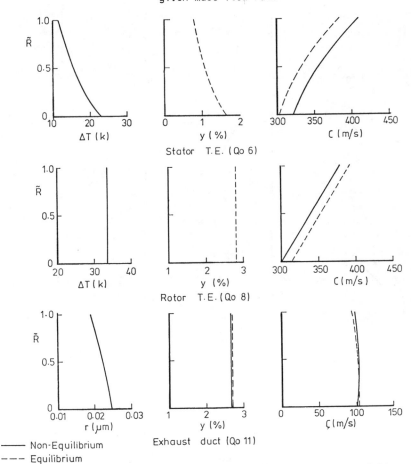

FIGURE 4.1.15 Theoretical results for the turbine stage of figure 4.1.12 for a prescribed stage pressure ratio

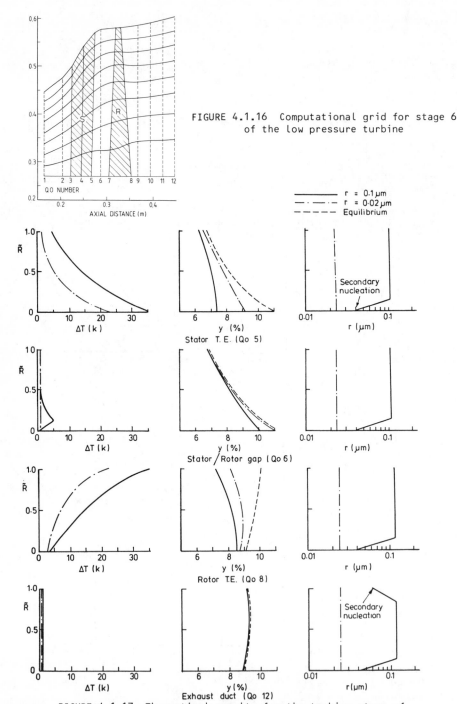

FIGURE 4.1.16 Computational grid for stage 6
of the low pressure turbine

FIGURE 4.1.17 Theoretical results for the turbine stage of
figure 4.1.16 for different droplet radii

for heat transfer. In contrast to the stator, the maximum supercooling occurs at the tip of the blade and, for droplets of radius 0.1 µm or larger, this is sufficient to promote a secondary nucleation.

On emerging from the rotor into the exhaust duct, equilibrium is rapidly re-established and by Q012 both the supercooling and the deviation from equilibrium wetness have been reduced almost to zero for all droplet sizes. The overall results of the calculations are summarized in Table 4.1.1.

	Equilibrium	Non-Equilibrium		
r_{inlet} (µm)		0.02	0.1	0.2
p_{inlet} (bar)	0.157	0.157	0.157	0.157
y_{inlet} (%)	5.4	5.4	5.4	5.4
p_{exit} (bar)	0.044	0.044	0.044	0.044
ϕ	1.0	1.018	1.048	1.063
Δs_T (kJ/kg K)		0.006	0.013	0.016
Δs_A (kJ/kg K)	0.094	0.094	0.094	0.094
Δh_t (kJ/kg)	133.1	132.2	131.1	130.1
Power (kW)	586.6	593.3	600.8	610.0
η_{TT} (%)	82.2	81.3	80.2	79.7

TABLE 4.1.1 Overall results of the calculations
for LP stage in figure 4.1.16

Of particular interest is the thermodynamic entropy rise Δs_T which increases with increasing droplet radius. However, the loss due to secondary nucleation has probably been underestimated by the method of averaging used and no account has been taken of other losses due to the mechanical effects of wetness. The mass flow coefficient increases with droplet radius and the stage enthalpy drop Δh_t decreases due to the increased thermodynamic loss. The power output, being the product of mass flow rate and Δh_t displays a slight increase, but whether this is true in general depends on whether the increase in mass flow rate with droplet size outweighs the reduction in Δh_t due to increased wetness loss.

4.1.4.5 The Ansaldo 320 MW, LP Turbine

This turbine has been the subject of a number of theoretical and experimental investigations and full details can be found in [4.24]. The LP cylinder has six stages, but for computational purposes it was only necessary to consider the flow in the final two stages as the steam remains superheated during its passage through the first four. The computational quasi-orthogonals are shown in figure 4.1.18 and the aerodynamic loss coefficients in figure 4.1.19. The latter were obtained from an equilibrium calculation and account for the profile, secondary and tip-leakage losses.

Table 4.1.2 shows the overall results for the two stages as a comparison between equilibrium and non-equilibrium solutions. This is a more useful presentation of the results than a direct comparison with the experimental data, because it highlights the differences caused by departures from equilibrium. It can be seen that there is an increase in mass flow rate of 3.1% and a reduction in stagnation enthalpy drop of 2.2%, giving an overall increase in power output of 0.9%. The thermodynamic wetness loss reduces the average total-total

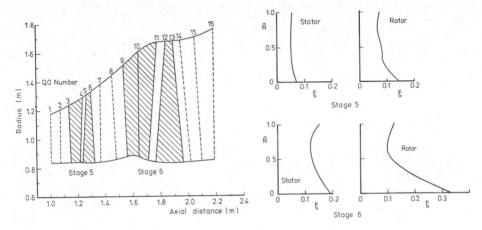

FIGURE 4.1.18 Computational quasi-   FIGURE 4.1.19 Aerodynamic loss
orthogonals for stages 5 and 6 of the coefficients for the Ansaldo turbine
Ansaldo 320 MW turbine

	Equilibrium	Non-Equilibrium
P_{inlet}(bar)	0.753	0.753
T_{inlet}(K)	370.5	370.5
P_{exit}(bar)	0.085	0.085
ϕ	1.0	1.031
Δs_T (kJ/kg K)		0.015
Δs_A (kJ/kg K)	0.102	0.102
Δh_t (kJ/kg)	273.5	267.6
Power (MW)	24.95	25.17
η_{TT} (%)	89.4	87.5

TABLE 4.1.2 Overall results of the calculations
for Ansaldo 320 MW LP turbine in figure 4.1.18

isentropic efficiency for the two stages by 1.9%, but other sources of loss due
to wetness have not been included and may be significant.
    Departure from equilibrium and the formation and growth of the liquid
in stage 5 are shown in figure 4.1.20. Nucleation first occurs at the trailing
edge of the stator blade near the hub (Q04) and the droplets grow rapidly in the
inter-row gap. By the rotor inlet (Q05), equilibrium has been completely re-
established in the hub streamtube, although the remainder of the flow is still
dry and supercooled. The balance is redressed during the expansion through the
rotor and by the leading edge of stage 6 stator (Q09) reversion over the whole
flow field is complete. Note, however, the considerable variation of droplet
radius across the annulus.

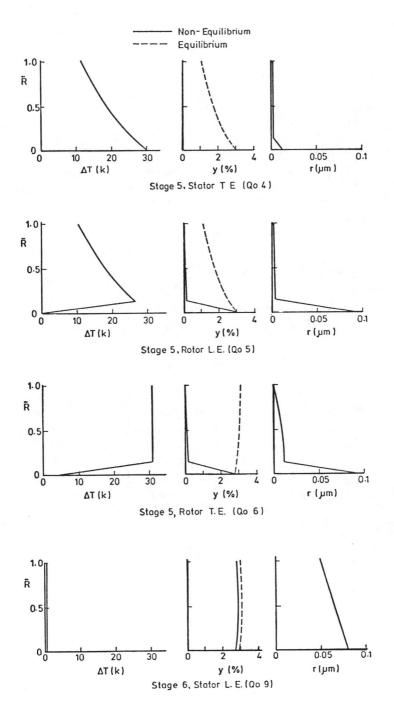

FIGURE 4.1.20 Supercooling and wetness in stage 5 of the Ansaldo turbine

J.B. Young

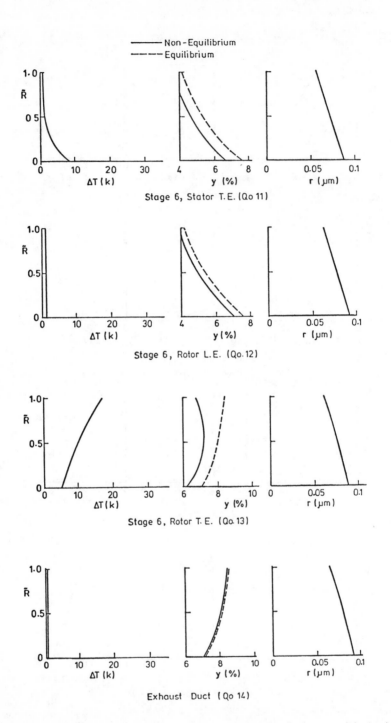

FIGURE 4.1.21   Supercooling and wetness in stage 6 of the Ansaldo turbine

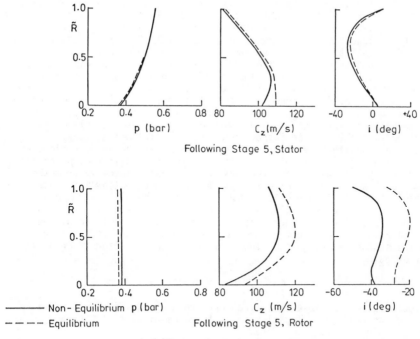

Following Stage 5, Stator

——— Non- Equilibrium p (bar)

— — — Equilibrium

Following Stage 5, Rotor

FIGURE 4.1.22   Variation of aerodynamic parameters
in stage 5 of the Ansaldo turbine

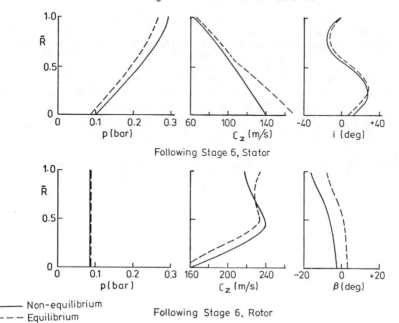

Following Stage 6, Stator

Following Stage 6, Rotor

——— Non-equilibrium

— — — Equilibrium

FIGURE 4.1.23   Variation of aerodynamic parameters
in stage 6 of the Ansaldo turbine

The results for stage 6 appear in figure 4.1.21. Due to the high rate of
expansion the vapour supercools, but never reaches a level sufficient to promote
secondary nucleation. This is attributable mainly to the fine droplet size of
the fog. At the leading edge of the rotor (QO12), the steam is in equilibrium
despite the fact that the equilibrium and non-equilibrium wetness fractions are
not identical. This is caused by the differences in pressure predicted by the
two solutions at this station. The final droplet radius at the exhaust varies
from 0.09 μm at the hub to 0.07 μm at the casing.
    Figures 4.1.22 and 4.1.23 demonstrate the differences between the equili-
brium and non-equilibrium analysis for some aerodynamic parameters of interest.
The spanwise variation of pressure and axial velocity after each blade row is
shown together with the flow incidence angle onto the following blade.
Figure 4.1.22 deals with stage 5 and it can be seen that the aerodynamic operation of the stator is only slightly affected by the non-equilibrium phenomena.
The axial velocity following the rotor, however, deviates markedly from the
equilibrium value and gives rise to a significant change in flow incidence onto
the stator of stage 6. Figure 4.1.22 shows that the non-equilibrium solution
predicts incidence angles some 10° more negative than the equilibrium result.
    Similar information for stage 6 can be found in 4.1.23. It should be noted
that there is a considerable difference between the two pressure distributions
following the stator, although the flow incidence onto the rotor is only
marginally affected.

4.1.4.6 Conclusions

1. Supercooled steam is denser than the corresponding equilibrium mixture and
has the effect of reducing velocities in LP turbines by up to 10% of the values
calculated assuming equilibrium flow.
2. Departures from equilibrium can increase the swallowing capacity of a turbine
stage by up to 6%. The mass flow coefficient is a strong function of droplet
size.
3. The pressure distribution and stream surfaces in LP turbines are only marginally affected by non-equilibrium effects. It is therefore unlikely that
supercooling of the flow could be detected by static pressure measurements.
4. In regions of high supercooling, changes of up to 10° in the relative inlet
flow angle onto the blades have been predicted. Generally the effect is to make
the flow incidence more negative compared with the equilibrium value.
5. Considerable spanwise variation of droplet size may occur due to the varying
locations of the Wilson Point on different stream surfaces.
6. Secondary nucleations are possible in the last stage of an LP turbine and
are controlled by the size of the primary fog and the local expansion rate. The
most likely locations are the hub region of the stator blade and the tip region
of the rotor.
7. The stage thermodynamic loss does not vary linearly with the mean wetness
fraction as suggested by the Baumann rule. Typically its contribution can
account for a reduction in efficiency of up to 7% in the primary nucleating
stage. For subsequent stages its magnitude depends on the prevailing droplet
size. Losses due to the mechanical effects of wetness have not been calculated
but may be significant for larger droplets.
8. The computational time for the non-equilibrium analysis of the last two
stages of the Ansaldo turbine was 13 minutes on an IBM 370/165 with a good opti-
mizing compiler. Although this is considerably greater than the 1 minute taken
for the equivalent equilibrium solution, there is still much scope for improving
the computational efficiency of the method

4.1.5 Closure

The theory of non-equilibrium wet steam flows has now reached a stage of development where it can be applied to two dimensional calculations in steam turbines.

Experimental verification of the theory is difficult, but is essential if such calculations are to be of value for turbine design. The most promising techniques are direct measurements of (a) the fog droplet size by light attenuation techniques, (b) the coarse water distribution using 'catchpot' probes and (c) the overall wetness fraction using total enthalpy probes. Much more data are required from all three sources, but the rate of accumulation is low. At present, improvements in turbine performance stem from a better understanding of the flow aerodynamics, but are subject to the law of diminishing returns. In the future, increases in turbine efficiency may well result from a better appreciation of thermodynamic phenomena, but, as yet, such possibilities remain virtually untapped.

4.2 TWO DIMENSIONAL CONDENSING FLOW IN TRANSONIC TURBINE CASCADES

J. Snoeck

4.2.1 Introduction

Numerous publications deal with the prediction of spontaneous condensation in nozzles. Because of the numerical schemes that are used, most of these methods are restricted to purely supersonic flows [4.25,4.26,4.27,4.28,4.29]. A more general algorithm is proposed in [4.30], which enables aerothermal shocks to be treated but the resulting procedure becomes rather complex. A general survey of the main features concerning one dimensional condensing flows is given by Gyarmathy and Moore [4.3].
    Although a considerable amount of work has been devoted to nozzle flows, very few results can be applied to the design of condensing turbines. Bakhtar and Mohammadi Tochai [4.31] were the first to present an approach for the two dimensional condensing flows through transonic cascades. They solve the conservation equations for the vapour-liquid mixture with a time marching finite volume technique.
    In the approach presented here, the conservation laws are formulated separately for each phase. After some simplifications, the vapour phase equations are integrated with a time marching scheme while an analytical solution of the liquid phase equations provides the droplet growth model. Some aspects were already presented in [4.32] and complete information can be found in [4.33].

4.2.2 Mathematical Description of the Flow

The conservation laws of multiphase systems can be expressed under several forms according to the parameter describing the proportion of each phase : mass fraction, volume fraction, molar concentration. In this context, the wetness rate y has been chosen, that is the ratio between the mass of the liquid phase and the mass of the whole system. The relationships have been written with the following assumptions :
A1 - The vapour phase is not viscous and not heat-conducting except for the interactions with the liquid phase.
A2 - The liquid phase is described by myriads of droplets that are considered as rigid spheres without rotational motion nor internal circulation of mass. The droplets do not interact between themselves; this means that they do not carry any electrical charge, that they do not coagulate nor break up.
    The work required to extend the droplet surface is negligible compared to the heat released by the condensation.

A3 - The influence of the walls under the form of mass deposition, viscous forces and heat transfer is ignored.
A4 - No external force acts upon the system; no chemical reaction occurs in it and no energy is exchanged by radiation. Finally, no mechanical work is extracted from the system or provided to it.

4.2.2.1 General Formulation of Conservation Laws

The time derivative of the liquid phase mass contained in a volume v of the flow is equal to the formation of liquid by nucleation and by the growth of the existing droplets.
    The total mass of the liquid phase is obtained by summing up the mass of each individual droplet. Assuming that the droplets are described by a finite number of classes defined by a radius r_i and a number n_i, the conservation law can be written under the form :

$$\frac{d}{dt}\left[\int_v \rho \left(\sum_i n_i \frac{4\pi r_i^3}{3} \rho_\ell\right) dv\right] = \int_v \rho \left(1 - \sum_i n_i \frac{4\pi r_i^3}{3} \rho_\ell\right) J \frac{4\pi r_\star^3}{3} \rho_\ell \, dv$$

$$+ \int_v \rho \left[\sum_i n_i \, 4\pi r_i^2 \left(\frac{dr_i}{dt}\right) \rho_\ell\right] dv \tag{4.2.1}$$

The definitions :

$$y \overset{\Delta}{=} \sum_i n_i \frac{4\pi r_i^3}{3} \rho_\ell \tag{4.2.2}$$

and

$$\mu_\ell \overset{\Delta}{=} (1-y) \, J \, \frac{4\pi r_\star^3}{3} \rho_\ell + \sum_i n_i \, 4\pi r_i^2 \left(\frac{dr_i}{dt}\right) \rho_\ell \tag{4.2.3}$$

allow us to deduce a more compact form of equation (4.2.1), that is :

$$\frac{d}{dt}\left[\int_v \rho y \, dv\right] = \int_v \rho \mu_\ell \, dv. \tag{4.2.4}$$

    The droplets have different velocities and thermodynamic states. However, because of their large number, it is possible to treat them as a continuum by the use of statistical mean values [4.34]. Let us consider the variable φ_i that describes some specific property of the droplets with a radius r_i. The appropriate mean value φ to be used in the conservation equations would be :

$$\varphi = \frac{\sum_i n_i \frac{4\pi r_i^3}{3} \rho_\ell \varphi_i}{\sum_i n_i \frac{4\pi r_i^3}{3} \rho_\ell}. \tag{4.2.5}$$

The theorems of Leibniz and Gauss-Ostrogradski can then be applied to equation (4.2.4) to get :

$$\frac{\partial}{\partial t}(\rho y) + \nabla \cdot \left(\rho y \vec{C}_\ell\right) = \rho \mu_\ell. \qquad (4.2.6)$$

The <u>continuity equation for the vapour phase</u> is derived in a similar way and is expressed as :

$$\frac{\partial}{\partial t}\left[\rho(1-y)\right] + \nabla \cdot \left[\rho(1-y)\vec{C}_g\right] = -\rho\mu_\ell \qquad (4.2.7)$$

The time derivative of the <u>momentum quantity of the liquid phase</u> contained in a volume v of the flow is equal to the sum of the forces acting on the droplets to which must be added the momentum transfer associated with the mass transfer μ_ℓ. The result of the derivation is given by the equation :

$$\frac{\partial}{\partial t}(\rho y \vec{C}_\ell) + \nabla \cdot \left(\rho y \vec{C}_\ell \otimes \vec{C}_\ell\right) = \rho\left(\vec{P}_\ell + \vec{D}_\ell + \mu_\ell \vec{C}_t\right) \qquad (4.2.8)$$

where $\vec{P}_\ell$ and $\vec{D}_\ell$ denote respectively the pressure and drag forces acting on the droplets contained in a mass unit of the mixture. $\vec{C}_t$ represents the mean velocity of the mass changing from one phase to the other one : its values lie between $\vec{C}_\ell$ and $\vec{C}_g$ and depend on the conditions in which the mass transfer occurs. The value of $\vec{P}_\ell$ can be obtained from the integration of the pressure forces acting on an isolated droplet and is given by the relationship :

$$\vec{P}_\ell = -\frac{y}{\rho_\ell}\nabla p. \qquad (4.2.9)$$

The value of $\vec{D}_\ell$ can be deduced from the usual drag laws.

The time derivative of the <u>momentum quantity of the vapour phase</u> contained in a volume v of the flow is equal to the pressure forces acting on the volume minus the forces exerted on the droplets and the momentum quantity transferred to the liquid phase. This can be expressed by the equation :

$$\frac{\partial}{\partial t}\left[\rho(1-y)\vec{C}_g\right] + \nabla \cdot \left[\rho(1-y)\vec{C}_g \otimes \vec{C}_g\right] = -\nabla p - \rho\left[\vec{P}_\ell + \vec{D}_\ell + \mu_\ell \vec{C}_t\right], \qquad (4.2.10)$$

The time derivative of the <u>total energy of the liquid phase contained</u> in a volume v of the flow is equal to the power of the pressure and drag forces exerted on the droplets, plus the heat transfer $\dot{q}_\ell$ due to the temperature difference between the phases and the transfer of total enthalpy associated with the mass transfer, i.e.,

$$\frac{\partial}{\partial t}(\rho y E_\ell) + \nabla \cdot \left(\rho y E_\ell \vec{C}_\ell\right) = \rho\left(\vec{P}_\ell \cdot \vec{C}_\ell + \vec{D}_\ell \cdot \vec{C}_\ell + \dot{q}_\ell + \mu_\ell h_t\right) \qquad (4.2.11)$$

where h_t denotes the mean specific total enthalpy of the transferred mass.

The time derivative of the <u>total energy of the vapour phase</u> contained in a volume v of the flow is equal to the power of the pressure forces exerted on the volume minus (a) the power of the forces acting on the droplets and (b) the transfer of heat and total enthalpy to the liquid phase. As the vapour phase does not occupy the entire volume v and since the droplets are considered as rigid spheres, the power of the pressure forces exerted on the volume v must be reduced by the factor $\rho \left(\dfrac{1-y}{\rho_g} \right)$. The final result can be written in the form :

$$\frac{\partial}{\partial t} \left[\rho(1-y)E_g \right] + \nabla \cdot \left[\rho(1-y)E_g \vec{C}_g \right] = - \rho \left(\frac{1-y}{\rho_g} \right) \nabla \cdot (p\vec{C}_g) - \rho \left(\vec{p}_\ell \cdot \vec{C}_\ell + \vec{D}_\ell \cdot \vec{C}_\ell + \dot{q}_\ell + \mu_\ell h_t \right)$$

(4.2.12)

If the values of the mass, momentum and heat transfer are known, the solution of the system of equations (4.2.6 to 4.2.12) is entirely determined.

<u>Simplifying assumptions</u>

Some further simplifications are made.
The volume of the liquid phase can be neglected compared to that of the vapour phase. This leads to the equation :

$$\rho_g = \rho(1-y).$$

(4.2.13)

Hence, the term $\rho\vec{p}_\ell$ can be neglected with respect to the quantity, ∇p, because it follows from equation (4.2.13) :

$$\nabla p + \rho\vec{p}_\ell = \rho\nabla p \left(\frac{1-y}{\rho_g} \right) \cong \nabla p.$$

(4.2.14)

The second assumption imposes equal velocities for both phases. This is acceptable for droplet radii smaller than 1 μm as encountered in the fog generated by homogeneous nucleation. This is expressed by the equality :

$$\vec{C}_\ell = \vec{C}_g = \vec{C}_t,$$

(4.2.15)

that implies :

$$\vec{D}_\ell = \vec{0}.$$

(4.2.16)

4.2.2.2 *System of Equations*

<u>Vapour phase</u>

With the simplifications proposed in the preceding section, the vapour phase equations expressing the conservation of mass, momentum and energy can be formulated as :

$$\frac{\partial}{\partial t} (\rho_g) + \nabla \cdot \left[\rho_g \vec{C}_g \right] = - \rho u_\ell$$

(4.2.17)

$$\frac{\partial}{\partial t} \left[\rho_g \vec{C}_g \right] + \nabla \cdot \left[\rho_g \vec{C}_g \otimes \vec{C}_g + p \right] = - \rho u_\ell \vec{C}_g$$

(4.2.18)

$$\frac{\partial}{\partial t} (\rho_g E_g) + \nabla \cdot \left[\rho_g E_g \vec{C}_g + p\vec{C}_g \right] = - \rho \dot{q}_\ell - \rho u_\ell h_t$$

(4.2.19)

The solution of the above set of equations (4.2.17), (4.2.18) and (4.2.19) requires the knowledge of the steam properties : equation of state, internal energy and entropy.

The equation of state is the truncated virial form :

$$p = \rho_g R T_g (1+B\rho_g) \tag{4.2.20}$$

proposed in [4.35] as well as an expression for the specific heat capacity at zero pressure.

The saturation line is defined from the formulas presented in [4.35] and [4.36] while the dynamic viscosity of steam and its thermal conductivity are estimated from the dry saturated values [4.37]. The expression of Grigull and Bach has been retained for the surface tension of water [4.38].

Liquid phase

Because the volume occupied by the droplets has been assumed to be negligible, the mass of the liquid phase cannot be deduced anymore from the continuity equation (4.2.6). Hence, this information will be provided by the classical nucleation theory and by the model of Gyarmathy for the droplet growth.

The nucleation rate is computed by the classical theory corrected for the non-isothermal effects as proposed by Kantrowitz [4.6]. More information about this aspect is given by Young in chapter 4.1.

It should be mentioned here that the classical theory has been throughly revised by Hedbäck [4.39]. Keeping the same basic ideas, this author proposes a rigorous derivation that accounts for all the aspects including the compressibility of the liquid phase. Unfortunately, this approach leads to time consuming solving procedures and could not be used for this reason.

The droplet growth model can be deduced from the momentum and energy equations of the liquid phase. The result of this derivative for an isolated droplet is given by :

$$\frac{4\pi r^3}{3} \rho_\ell \frac{de_\ell}{dt} = \dot{q}_d + \mu_d(h_g - e_\ell) \tag{4.2.21}$$

where the left hand side represents the heat capacity of the droplet. This term is small compared to the heat release due to the condensation process and is usually neglected. Replacing the difference $(h_g - e_\ell)$ by the enthalpy difference $(h_g - h_\ell)$ provides the model for the droplet growth :

$$\dot{q}_d + \mu_d (h_g - h_\ell) = 0. \tag{4.2.22}$$

The heat transfer to the droplet can be expressed as :

$$\dot{q}_d = 2\pi r \lambda_g (T_g - T_r) \, Nu_H \tag{4.2.23}$$

where the Nusselt number Nu_H is estimated from the multirange expression given by Gyarmathy in [4.40] :

$$Nu_H = \frac{2}{1+ \dfrac{2}{B_H} Kn} \tag{4.2.24}$$

with

$$B_H = \frac{\eta_g}{1.24\lambda_g}\left(c_{vg} + \frac{R}{2}\right) \tag{4.2.25}$$

and

$$Kn = \frac{\eta_g}{r\rho_g\sqrt{RT_g}} \tag{4.2.26}$$

The droplet surface temperature T_r can be estimated by the following relationship given by the same author :

$$T_r = T_g + \Delta T\left(1 - \frac{r_\star}{r}\right). \tag{4.2.27}$$

Estimating the enthalpy of the liquid phase h_ℓ by the formula [4.31].

$$h_\ell = c_{p\ell}(T_r - T_D) \tag{4.2.28}$$

allows to integrate the equation (4.2.22) for constant vapour conditions with the initial values :

$$r = r_0 \qquad \text{and} \qquad t = t_0$$

to get the form :

$$\left(h_g - c_{p\ell}(T_g - T_D)\right)\left(1 + \frac{2Kn_\star}{B_H}\right)\ell n\frac{z-1}{z_0-1} + \left[\left(h_g - c_{p\ell}(T_s - T_D)\right)\left(1 + \frac{2Kn_\star}{B_H}\right) + c_{p\ell}(T_s - T_g)\right](z - z_0)$$

$$+ \left(\frac{1}{2}h_g - c_{p\ell}(T_s - T_D)\right)(z^2 - z_0^2) = \frac{\lambda_g\Delta T}{\rho_\ell r_\star^2}(t - t_0) \tag{4.2.29}$$

where z represents the adimensional radius defined as :

$$z = \frac{r}{r_\star}. \tag{4.2.30}$$

4.2.3 Condensing Flows in Transonic Cascades

4.2.3.1 *Calculation Procedure*

The calculation procedure combines a time marching method for the vapour phase equations with an analytical treatment of the liquid phase equations. A general layout of the algorithm is shown in figure 4.2.1. From the vapour conditions prevailing at time t, the variables of the liquid phase can be determined with the nucleation theory and the droplet growth model. The transfer rates are then known as well as the sink and source terms appearing at the right hand side of the vapour phase equations. This set can now be integrated from time t to time t+Δt to update the variables of the vapour phase and the algorithm goes on until the steady state solution is obtained.

Vapour phase

The physical domain of the flow is defined in figure 4.2.2. It is constituted by one blade passage extended upstream and downstream in the approximate direction of the flow over a length which is of the order of one chord. The lines AB, CD, EF, GH, are periodic boundaries while AE and DH represent the inlet plane and the outlet plane respectively. The suction side FG and pressure side BC are impermeable boundaries across which no mass or energy tranfer can occur; hence, the transport of momentum is restricted to the forces exerted by the static pressure on the walls.

The domain is discretized by equidistant pseudostreamlines and pitchwise lines. The location of the pitchwise lines is chosen to concentrate the computational effort in the regions where the highest gradients appear.

The nodes of the mesh are defined by the intersections of these two families of curves. The control surface is the bitrapezoidal element developed by Van Hove & Arts (Fig. 4.2.3). It has shown to fulfil the criteria of consistency, accuracy and computational time [4.41].

In a two dimensional geometry, the equations (4.2.17) and (4.2.19) can be represented under the vectorial form :

$$\frac{\partial \vec{s}}{\partial t} + \frac{\partial \vec{f}_1}{\partial x_1} + \frac{\partial \vec{f}_2}{\partial x_2} = \vec{h} \tag{4.2.31}$$

where the subscripts 1 and 2 refer respectively to the axial and tangential directions. $\vec{s}$ is the vector of the unknowns and $\vec{f}_j$ denotes the vector of the convective terms related to the direction Ox_j.

Let us consider now the control surfaces S and its contour ∂S on which the outgoing normal $\vec{n}$ has been defined with its components n_1 and n_2. Inside the contour the values of the unknowns are supposed to be constant. Applying then the theorem of Gauss to the space derivatives allows the equation (4.2.31) to be integrated to give :

$$S \frac{\partial \vec{s}}{\partial t} = - \int_{\partial S} (\vec{f}_1 n_1 + \vec{f}_2 n_2) \cdot \vec{d\ell} + S\vec{h}. \tag{4.2.32}$$

This equation states that the variation of $\vec{s}$ inside the control surface S is equal to the net convective transport of $\vec{s}$ through the boundary ∂S to which must be added the contribution of the local sources.

I_1 and I_2 being the indices related to the axial and tangential directions, the time derivative of equation (4.2.32) is then discretized following the time marching scheme originally proposed by McDonald [4.43] . The result is expressed as :

$$\vec{s}_{I_1,I_2}^{t+\Delta t} = \frac{1}{4} \left[\vec{s}_{I_1-1,I_2}^{t+\Delta t} + \vec{s}_{I_1+1,I_2}^{t} + \vec{s}_{I_1,I_2-1}^{t} + \vec{s}_{I_1,I_2+1}^{t} \right]$$

$$- \frac{\alpha}{4} \left[\vec{s}_{I_1-1,I_2}^{t_0} + \vec{s}_{I_1+1,I_2}^{t_0} + \vec{s}_{I_1,I_2-1}^{t_0} + \vec{s}_{I_1,I_2+1}^{t_0} - 4\vec{s}_{I_1,I_2}^{t_0} \right]$$

$$+ \frac{\Delta t}{S} \Sigma \text{ (transport terms)}^t + \Delta t \, \vec{h}_{I_1,I_2}^{t} \tag{4.2.33}$$

where the terms superscripted by t_0 are updated every Nv iterations. The numerical viscosity coefficient α is proportional to the discretized second derivative of the density and its value is given by

$$\alpha = Vc \left[1 - \frac{\left| (\rho_g)^{t_0}_{I_1-1,I_2} + (\rho_g)^{t_0}_{I_1+1,I_2} + (\rho_g)^{t_0}_{I_1,I_2-1} + (\rho_g)^{t_0}_{I_1,I_2+1} - 4(\rho_g)^{t_0}_{I_1,I_2} \right|}{4} \right] \quad (4.2.34)$$

where Vc is a parameter near 1. It is chosen empirically.

The stability condition of an explicit scheme is given by the C.F.L. criterion. In the case of the bitrapezoidal element, this condition is expressed as:

$$\left(\| \vec{C} \| + a_f \right) \frac{\Delta t}{\Delta \ell} \leqslant 1 \quad (4.2.35)$$

where $\Delta \ell$ denotes the shortest among the distances from the node (I_1,I_2) to the four sides of the quadrangle defined by the nodes (I_1-1,I_2), (I_1+1,I_2), (I_1,I_2-1), (I_1-I_2+1). a_f represents the frozen speed of sound, that is the highest velocity at which information is transmitted through the fluid. Its value calculated for the truncated virial equation of state (4.2.20) is given by :

$$\frac{a_f^2}{RT_g} = 1 + 2B\rho_g + \frac{R\left[1+(B+T_g\dot{B})\rho_g \right]\left[1+(B-T_g\dot{B})\rho_g \right]}{c_{vgo}(T_g) - (2\dot{B}+T_g\ddot{B})\rho_g RT_g} \quad (4.2.36)$$

with

$$\dot{B} = \frac{dB}{dT_g} \quad (4.2.37)$$

and

$$\ddot{B} = \frac{d^2B}{dT_g^2}. \quad (4.2.38)$$

With the scheme (4.2.33) the vapour phase equations can be integrated in all the nodes located inside the domain as well as along the periodic boundaries. An excentered numerical scheme derived from the centered version (4.2.33) is applied to the nodes along the impermeable boundaries.

The theory of characteristics allows the number of physical conditions to be determined that must be imposed along the pitchwise boundaries. The axial flow being assumed to be subsonic in both cases, the number of those conditions is equal to 3 in the inlet plane and to 1 in the outlet plane. The boundary conditions have been chosen following the recommendations of [4.41].

In the inlet plane, the flow is assumed to be dry and the total enthalpy and entropy to be equal to the values of the upstream total conditions. The third physical condition is given by the inlet flow angle. As the number of these conditions (3) is smaller than the number of unknowns (4), the additional information must be provided from the computational domain under the form of a numerical condition. Therefore, the axial momentum equation is integrated with a downstream excentered scheme.

In the outlet plane, the physical condition is the static pressure. The extra information results from the integration of the conservation equations for the mass, the axial momentum and the tangential momentum with an upstream excentered scheme.

Liquid_phase

The appearance and the growth of the liquid phase through the cascade depends on the flow history and this computation requires a knowledge of the trajectories of the droplets. As these droplets are very small, the velocities of both phases have been assumed to be equal and the problem is reduced to the calculation of the vapour phase streamlines.

Let us consider the nodal point I_1, I_2 in figure 4.2.4. The streamline passing through it intersects the plane $I_1 - \frac{1}{2}$ in the point $P(I_1, I_2)$ and the plane $I_1 + \frac{1}{2}$ in the point $Q(I_1, I_2)$. The growth of the existing droplets and the birth of the new nuclei along the line $P(I_1, I_2)$ $Q(I_1, I_2)$ are evaluated with the thermodynamic conditions prevailing at the node (I_1, I_2). This allows all the unknowns at the station $I_1 + \frac{1}{2}$ to be deduced, enabling the calculation to proceed further downstream.

For the calculation of the location of the points $P(I_1, I_2)$ and $Q(I_1, I_2)$, let us consider the quadrangle ABCD in figure 4.2.5. A is the origin of a cartesian coordinate system (x_1', x_2') pointing in axial and tangential directions respectively. Approximating locally the streamline AQ by a straight line allows us to write :

$$(x_2')_Q = \Delta \ell_1 \tan\beta \qquad\qquad (4.2.39)$$

where the mean value of the angle β between the streamlines and the x_1'-axis is estimated from the axial and tangential momentum quantities in the points A and Q.

An explicit equation can be found when writing the relationship (4.2.39) in the form :

$$(x_2')_Q = \Delta \ell + \xi_2 \Delta \ell_2 \qquad\qquad (4.2.40)$$

where $\Delta \ell$ is the distance from the point S to the x_1'-axis and $\Delta \ell_2$ the distance between N and S. ξ_2 now becomes the unknown that is adimensionalized by the length $\Delta \ell_2$. Hence, the momentum quantities in Q can be interpolated as a function of ξ_2 between the values in the points N and S and an explicit equation can then be derived for ξ_2 which takes the form :

$$(C_{1N} - C_{1S})\xi_2^2 + \left[(C_{1N} - C_{1S}) \frac{\Delta \ell}{\Delta \ell_2} + (C_{1A} + C_{1S}) - (C_{2N} - C_{2S}) \frac{\Delta \ell_1}{\Delta \ell_2} \right] \xi_2$$

$$+ (C_{1A} + C_{1S}) \frac{\Delta \ell}{\Delta \ell_2} - (C_{2A} + C_{2S}) \frac{\Delta \ell_1}{\Delta \ell_2} = 0. \qquad\qquad (4.2.41)$$

The solution of this allows us to compute the density and the momentum quantities in point Q as well as the time needed by the flow to move from the point A to the point Q.

In order to save computer time, the complete droplet spectrum is not retained but is replaced by a bidispersion.

At each node, the liquid phase is described by two droplet classes. The first one represents the droplets that are born upstream and will be called the "equivalent mono-dispersion"; its growth along the streamline $P(I_1, I_2)$ $Q(I_1, I_2)$

is computed by applying the equation (4.2.29) with the thermodynamic conditions
prevailing at the node (I_1, I_2). The second class represents the droplets nucleated along the streamline $P(I_1, I_2)$ $Q(I_1, I_2)$.

The problem of defining an equivalent monodispersion at the point $P(I_1, I_2)$
has already been treated in [4.33] and [4.34]. The liquid phase is defined by two
droplet classes obtained by interpolating between the values in the points
$Q (I_1-1, I_2)$. A comparison between measured and calculated pressure distributions
in one dimensional nozzles has shown that satisfactory predictions can be obtained
by imposing the conservation of the number of droplets and of their global growth
rate, i.e.:

$$n = \sum_{i=1}^{2} n_i \qquad (4.2.42)$$

and

$$\bar{n} \ \bar{\dot{q}}_d = \sum_{i=1}^{2} n_i \ \dot{q}_{di}. \qquad (4.2.43)$$

The overbar refers to the values of the equivalent monodispersion. Combining
the conditions (4.2.42) and (4.2.43) leads to an equation for the radius $\bar{r}$:

$$\bar{r}(\bar{T}_r - \bar{T}_g) \ Nu_H(\bar{r}) = \frac{\sum\limits_{i} n_i r_i (T_{ri} - T_g) Nu_H(r_i)}{\sum\limits_{i} n_i} \underset{=}{\triangle} K \qquad (4.2.44)$$

and finally, through the formulas (4.2.24) to (4.2.27), to the adimensionalized
equation :

$$\bar{z}^2 - \left(1 + \frac{K}{2\triangle Tr_\star}\right) \bar{z} - \frac{K \ Kn_\star}{\triangle TB_H r_\star} = 0 \qquad (4.2.45)$$

of which the positive root gives the equivalent radius.

At low pressure, when the Knudsen number Kn tends to infinity, the Nusselt
number Nu_H becomes :

$$Nu_H = \frac{B_H}{Kn}. \qquad (4.2.46)$$

If one assumes then the same temperature for all the droplets, one finds
from equation (4.2.44) :

$$\bar{r}^2 = \frac{\sum\limits_{i} n_i r_i^2}{\sum\limits_{i} n_i} \hat{=} r_{20}. \qquad (4.2.47)$$

This form of a surface-averaged radius is used by several authors [4.31, 4.20].

At higher pressure, when the Knudsen number Kn tends to zero, the Nusselt
number Nu_H becomes :

$$Nu_H = 2 \qquad (4.2.48)$$

and the equivalent radius can be approximated by :

$$\bar{r} = \frac{\sum\limits_{i} n_i\, r_i}{\sum\limits_{i} n_i} \hat{=} r_{10} \tag{4.2.49}$$

4.2.3.2 *Sample Calculations*

The sample calculations are related to a rotor tip section of the last stage of a large steam turbine (Fig. 4.2.6). The computations were done with 45 pitchwise lines and 11 pseudo-streamlines.

To investigate the influence of condensation on the cascade performances, the upstream total pressure was kept equal to 1 bar while the upstream total temperature was changed by steps of 20 K. This was done for three pressure ratios : 0.6, 0.5 and 0.35. Figure 4.2.7 presents on a Mollier chart the isentropic expansion lines related to these conditions. Unlike the expansion conditions prevailing in a nozzle, those experienced by steam in a cascade vary significantly across the blade passage.

Let us assume that the inlet flow conditions and the overall expansion ratio for a given cascade lead to thermodynamic non equilibrium conditions within the blade passage. The question is to know where the first droplets will occur. The strong acceleration along the blade suction surface in general causes the nucleation to start there. From the suction side, the nucleation zone will extend progressively across the entire blade passage. The droplets follow the vapour streamlines and continue to grow on their way to the cascade exit.

When the condensation occurs, the heat release causes the suction side static pressure to rise above the corresponding values of the superheated steam (Figs. 4.2.8a,b,c). The effect of the condensation on the blade pressure distribution becomes particularly evident when the condensing flow moves across a region where an aerodynamic shock is observed in dry conditions. Under these conditions, the effect of the condensation is to reduce gradually the shock intensity, but this happens by limiting the pressure drop before the shock interference point on the blade surface rather than the pressure rise after the shock.

The pressure distribution along the pressure side can also be affected where no local condensation is observed. A possible explanation is that for a given pressure ratio, the volume flow rate is not greatly modified with decreasing upstream total temperatures while the flow velocity along the suction side is significantly reduced. This involves a flow redistribution across the blade passage which is responsible for the pressure decrease observed in figures 4.2.8b,c.

Additional information is given in figures 4.2.9a,b,c by the wetness rate and subcooling distributions along the blade surface.

Downstream of the nucleation zone on the suction side, the rapid increase of the wetness rate involves an important heat release to the vapour phase and the subcooling falls off. This feature illustrates the natural tendency of the system to recover the thermodynamic equilibrium defined by the saturation conditions. These conditions are almost reached when the recompression through the trailing edge shock is weak and its position is far downstream from the region where the droplets appear and grow. In this case, the wetness fraction remains constant over a certain range, as seen for the pressure ratio 0.35 and the upstream total temperatures 373.15 K and 393.15 K (Fig. 4.2.9c).

When the temperature rise is sufficiently high through the recompression, a slightly superheated region can appear downstream and the liquid phase partially evaporates. This is observed for the pressure ratios 0.6 (Fig. 4.2.9a) and 0.5 with an upstream total temperature equal to 373.15 K (Fig. 4.2.9b).

In the other cases, the recompression through the trailing edge shock occurs at the place where the droplets would start to grow rapidly. Then, the rise of the vapour phase temperature affects the transfer rates and the wetness fraction

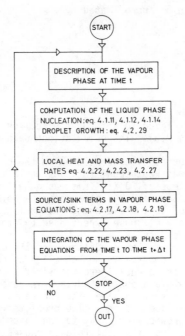

FIGURE 4.2.1 Algorithm for the computation

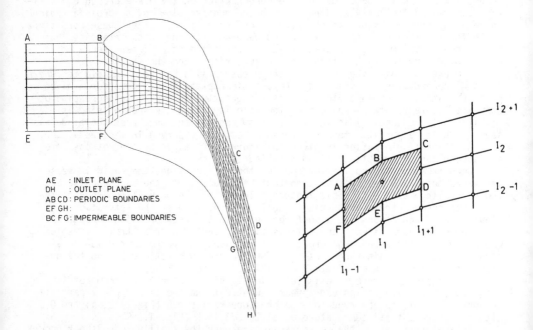

AE   : INLET PLANE
DH   : OUTLET PLANE
ABCD : PERIODIC BOUNDARIES
EF GH :
BC F G : IMPERMEABLE BOUNDARIES

FIGURE 4.2.2 Physical domain
of the flow

FIGURE 4.2.3 Bitrapezoidal element
of Van Hove & Arts

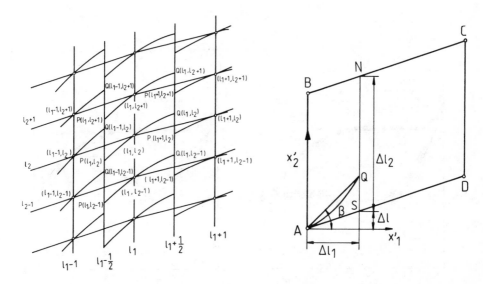

FIGURE 4.2.4 Definition of the streamlines   FIGURE 4.2.5 Calculation of a
                                                                    streamline

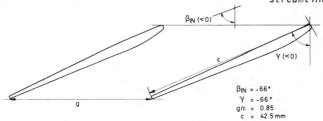

FIGURE 4.2.6 Rotor tip section of the last stage of a large steam turbine
             ([4.42 , blade 3])

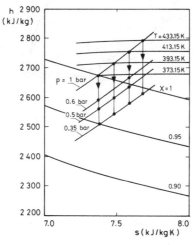

FIGURE 4.2.7 Isentropic expansion lines

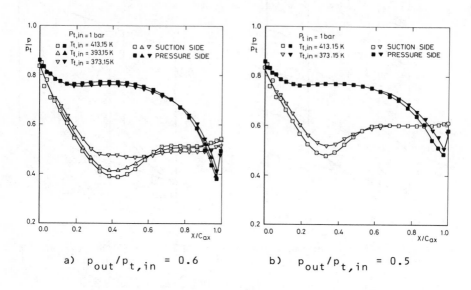

a) $p_{out}/p_{t,in} = 0.6$ b) $p_{out}/p_{t,in} = 0.5$

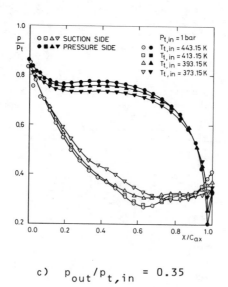

c) $p_{out}/p_{t,in} = 0.35$

FIGURE 4.2.8 Blade pressure distribution - influence of condensation

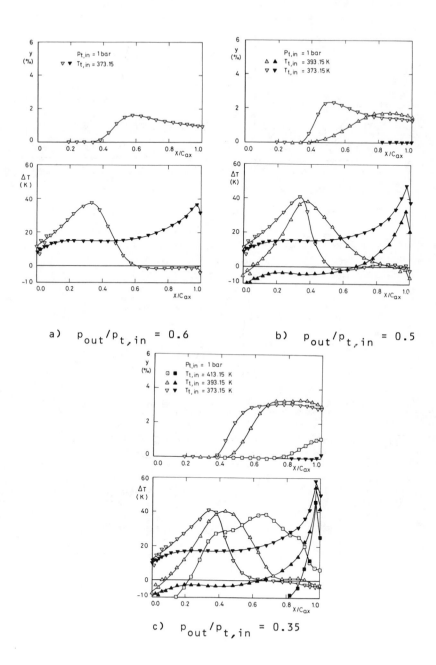

FIGURE 4.2.9 Blade wetnesss and subcooling distributions

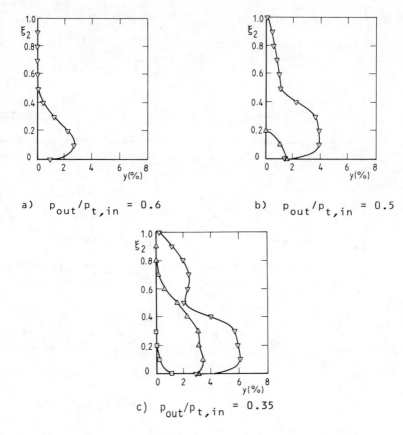

a) $p_{out}/p_{t,in} = 0.6$ b) $p_{out}/p_{t,in} = 0.5$

c) $p_{out}/p_{t,in} = 0.35$

FIGURE 4.2.10 Wetness distribution in the trailing edge plane

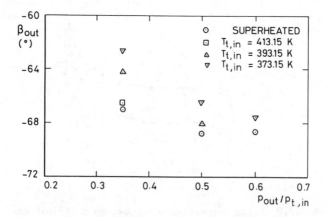

FIGURE 4.2.11 Outlet flow angle

grows more slowly. This situation is encountered for the pressure ratio 0.5 and an upstream total temperature 393.15 K, as well as for the pressure ratio 0.35 and the upstream total temperature 413.15 K (Fig. 4.2.9c).

The figures 4.2.10a,b,c show the wetness rate distribution in the trailing edge plane; the values $\xi_2 = 0$ and $\xi_2 = 1$ refer respectively to the positions of the suction and pressure sides. Due to the non uniform expansion through the cascase the major part of the liquid phase is contained in the lowest half of the passage.

Another effect of the condensation on the flow is shown in figure 4.2.11. The outlet flow angle defined with respect to the axial direction is given in function of the pressure ratio and the upstream total temperature. For a given pressure ratio, the flow direction becomes more axial as the wetness rate increases. The difference with the results obtained for superheated conditions can be as high as 4.5° for the pressure ratio 0.35.

4.2.4   Conclusions

Due to the lack of comparisons between predictions and experimental data, only tentative conclusions can be drawn. However, the calculations have pointed out some important aspects.
- The two dimensional character of the flow plays a decisive role in the appearance of the liquid phase. The first droplets are born along the suction side where the reversion to the thermodynamic equilibrium can take place before the flow reaches the trailing edge. This is not the case along the pressure side where the steam remains subcooled because of the lower expansion rate.
- The heat released by the droplets affects the blade pressure distribution. When an aerodynamic shock impinges on the suction side in dry conditions, its strength can be reduced substantially in the presence of condensation. The sample calculations have shown that the pressure distribution along the pressure side can also be modified.
- Finally, the outlet flow angle variation can differ by several degrees from the value computed for the same pressure ratio and superheated steam. This feature is due to the thermodynamic losses involved by the process and shows how much the cascade performances can be affected by condensation.

NOMENCLATURE

A	flow cross-sectional area
a_e, a_f	equilibrium and frozen speed of sound
B	virial coefficient
$\dot{B}$	$\dfrac{dB}{dT_g}$
$\ddot{B}$	$\dfrac{d^2B}{dT_g^2}$
b	blade blockage factor
C	velocity
C_m, C_θ	velocity in meridional and tangential directions
c	mixture specific heat capacity = $(1-y)c_{pg} + \Sigma y_i c_i$
c_{ax}	axial chord

c_{pg} isobaric specific heat capacity of vapour

c_v isochoric specific heat capacity

$\vec{D}_\ell$ total drag force exerted on the droplets per unit mass of mixture

$\dfrac{d}{dt}$ time derivative

E specific total energy $E = e + \dfrac{\vec{C} \cdot \vec{C}}{2}$

e specific internal energy

$\vec{f}_j$ vector of the convective terms related to the Ox_j-direction

h specific enthalpy of mixture

$\vec{h}$ vector constituted by the RHS terms of the conservation equations (4.2.31)

h_{fg} specific enthalpy of evaporation

h_g, h_i specific enthalpy of vapour, liquid

h_t stagnation specific enthalpy

I rothalpy

I_J nodal index in the Ox_j-direction

J nucleation rate per time and vapour mass unit

Kn Knudsen number

k Boltzmann constant ($1.380622 \ 10^{-23} J/K$)

k_e, k_f equilibrium and frozen isentropic exponents

ℓ distance

M Mach number or molar mass

m mass of a droplet or meridional direction

$\dot{m}$ mass flow rate

N_0 Avogadro's number

n number of droplets per unit mass of steam

$\vec{n}_i$ number of droplets with a radius r_i per unit mass of mixture

Pr_g Prandtl number of vapour

p vapour pressure

$\vec{P}_\ell$ pressure force exerted on the droplets per unit mass of mixture

p_t stagnation vapour pressure

Q angle between turbine axis and quasi-orthogonal

q quasi-orthogonal direction

q_c condensation coefficient

$\dot{q}_d$ heat transfer to the droplet per time unit

$\dot{q}_\ell$ heat transfer to the liquid phase per time unit and per unit mass of mixture

R gas constant (461.51 J.kgK)

r droplet radius

$r_\star$	Kelvin-Helmholtz critical radius
r_c	radius of curvature of streamsurface in meridional plane
r_s	radius of streamsurface from turbine axis
S	control surface
s	specific entropy of mixture
s_g, s_ℓ	specific entropy of vapour, liquid
T	temperature
T_D	datum temperature ($T_D = 273.16°K$)
T_g, T_i	vapour, liquid temperature
T_r	droplet surface temperature
T_s	saturation temperature
t	time
Vc	numerical viscosity parameter
v	control volume
x_j	coordinate in the $0x_j$-direction
y, y_i	wetness fraction
z	adimensionalized droplet radius, $z = \dfrac{r}{r_\star}$
α	numerical viscosity coefficient or coefficient of thermal expansion
β	flow angle (with respect to the axial direction), isothermal compressibility
∂S	boundary of the control surface S
$\dfrac{\partial}{\partial t}$	partial derivative with respect to time
$\Delta\ell, \Delta\ell_j$	distances measured in the cascade plane
$\Delta s_A, \Delta s_T$	aerodynamic and thermodynamic entropy increases
ΔT	vapour supercooling
$\Delta T_i, \Delta T_\ell$	capillary supercooling
Δt	time step
η_g	dynamic viscosity of steam
φ	meridional flow angle except if otherwise specified
ϕ	mass flow coefficient
λ_g	thermal conductivity of steam
μ_d	mass transfer to a droplet per time unit
μ_ℓ	mass transfer to the liquid phase per time unit and per mass unit of mixture
ξ_A, ξ_T	aerodynamic and thermodynamic energy loss coefficients
ξ_j	adimensional coordinate in the $0x_j$-direction
ρ	density
ρ_g, ρ_i	density of vapour, liquid

σ	surface tension
τ, τ_i	thermal relaxation time
Ω	rotational speed
$\hat{=}$	by definition
$\cdot$	scalar product
$\otimes$	tensor product
∇	nabla operator

Superscripts and subscripts

CL	given by the classical theory
g	of the vapour phase
i	related to the droplets class i
I_1, I_2	at the node (I_1, I_2)
j	related to the Ox_j-direction
ℓ	of the liquid phase
0	at zero pressure
s	saturation value
$\star$	at critical conditions
$\overline{}$	related to the equivalent monodispersion

REFERENCES

4.1 Gyarmathy, G.: Bases of a Theory for Wet Steam Turbines. *Bull. 6, Inst. for Thermal Turbomachines*, Federal Technical University, Zurich, 1962.
4.2 Kirillov, I.I. & Yablonik, R.M.: Fundamentals of the Theory of Turbines Operating on Wet Steam. *NASA Technical Translation* TT F-611, 1970.
4.3 Moore, K.J. & Sieverding, C. (Editors): Two-Phase Steam Flow in Turbines and Separators. *McGraw-Hill*, 1976.
4.4 Young, J.B.: The Spontaneous Condensation of Steam in Supersonic Nozzles. *Physico-Chemical Hydrodynamics*, Vol. 3, 1982, p 57.
4.5 McDonald, J.E.: Homogeneous Nucleation of Vapour Condensation. *Am. J. Phys.*, Vol. 30, 1962, p 870 and Vol. 31. 1963, p 31.
4.6 Kantrowitz, A.: Nucleation in very Rapid Vapour Expansions. *J. Chem. Phys.*, Vol. 19, 1951, p 1097.
4.7 Young, J.B.: Semi-Analytical Techniques for Investigating Thermal Non-Equilibrium Effects in Wet Steam Turbines. *Int. J. Heat & Fluid Flow*, Vol. 5, No. 2, June 1984, pp 81-91.
4.8 Stodola, A.: Steam and Gas Turbines. New York, *Peter Smith*, 1945.
4.9 Filippov, G.A.; Povarov, O.A.; Nickolsky, A.J.: The Steam Flow Discharge Coefficient and Losses in Nozzles of a Steam Turbine Stage Operating in the Low Steam Wetness Zone. In: *'Aero-Thermodynamics of Steam Turbines'*, ASME, 1981, pp 37-46.
4.10 Young, J.B.: Critical Conditions and the Choking Mass Flowrate in Non-Equilibrium Wet Steam Flows. *ASME Trans., Series I : Journal of Fluids Engineering*, Vol. 106, No. 4, December 1984, pp 452-458.
4.11 Bratuta, E.G.; Shatilov, S.P.; Pyasik, D.N.: A Comparison of the Throughput Capacities of the TS-1A and TS-1AV Blade Cascades with Flows of Superheated and Wet Steam. *Thermal Engineering*, Vol. 20, No. 2, February 1973, pp 117-121.
4.12 Deich, M.E.: Sound velocity and Decrement of Wave Attenuation in Two-Phase Media (in Russian). *Prace Inst. Maszyn Przeplywowych*, Poland, 1966, pp 29-31.
4.13 Konorski, A.: Propagation of Small Disturbances in Two-Phase Media (in Polish), *Prace Inst. Maszyn Przeplywowych*, Poland, Vol. 57, 1971, p 65
4.14 Petr, V.: Variation of Sound Velocity in Wet Steam. In: *'Wet Steam 4'*, Inst. Mech. Engrs., C24/73, 1973, p 17.
4.15 Konorski, A.: Critical Flow Conditions of Non-Equilibrium Two-Phase Media (in Polish), *Prace Inst. Maszyn Przeplywowych*, Poland, Vol. 65, 1974, p 3.
4.16 Konorski, A.: Critical Flow Conditions and Choking Flowrates of Wet Steam Flows. *Proc. of Seventh Conference on 'Steam Turbines of Large Output'*, Plzen, 1979, p 228.
4.17 Bakhtar, F. & Young, J.B.: A Study of Choking Conditions in the Flow of Wet Steam. *Proc. Instn. Mech. Engrs.*, Vol. 192, 1978, p 237.
4.18 Novak, R.A.: Streamline Curvature Computing Procedures for Fluid Flow Problems. *ASME Trans., Series A : Journal of Engineering for Power*, Vol. 89, No. 4, October 1967, pp 478-490.

4.19 Smith, L.H.: The Radial Equilibrium Equation of Turbomachinery. *ASME Trans., Series A : Journal of Engineering for Power*, Vol. 88, No. 1, January 1966, pp 1-12.

4.20 Yeoh, C.C. & Young, J.B.: Non Equilibrium Streamline Curvature Throughflow Calculations in Wet Steam Turbines. *ASME Trans., Series A : Journal of Engineering for Power*, Vol. 104, No. 2, April 1982, pp 489-496.

4.21 Yeoh, C.C. & Young, J.B.: Non-Equilibrium Throughflow Analyses of Low Pressure, Wet Steam Turbines. *ASME Trans., Series A : Journal for Gas Turbines and Power*, Vol. 106, No. 4, October 1984, pp 716-727.

4.22 Denton, J.D.: Throughflow Calculations for Transonic Axial Flow Turbines. *ASME Trans., Series A : Journal of Engineering for Power*, Vol. 100, No. 2, April 1978, pp 212-218.

4.23 Yeoh, C.C. & Young, J.B.: The Effect of Droplet Size on the Flow in the Last Stage of a One-Third Scale Model Low-Pressure Turbine. *Proc. Inst. Mech. Engrs.*, Vol. 198A, 1984, p 309.

4.24 Hirsch, C. & Denton, J.D. (Editors): Throughflow Calculations in Axial Turbomachines, *AGARD AR 175*, 1981.

4.25 Gyarmathy, G. & Meyer, H.: Spontane Kondensation. *VDI-Forschungsheft 508*.

4.26 Hill, P.G.: Condensation of Water Vapour During Supersonic Expansion in Nozzles. *Journal of Fluid Mechanics*, Vol. 25, Part 3, July 1966, pp 593-620.

4.27 Moore, M.J.; Walters, P.; Crane, R.; Davidson, B.: Predicting the Fog Drop Size in Wet Steam Turbines. *Inst. Mech. Engrs.*, CP 37/73, 1973, pp 101-109.

4.28 Bakhtar, F.; Riley, D.J.; Tubman, K.A.; Young, J.B.: Nucleation Studies in Flowing High Pressure Steam. *Proc. Inst. Mech. Engrs.*, Vol. 189, 41/75, pp 427-436.

4.29 Moses, C.A. & Stein, G.D.: On the Growth of Steam Droplets Formed in a Laval Nozzle Using Both Static Pressure and Light Scattering Measurements. *ASME Trans., Series I : Journal of Fluids Engineering*, Vol. 100, No. 3, September 1978, pp 311-322.

4.30 Barschdorff, D.: Verlauf der Zustandsgrössen und gasdynamische Zusammenhänge bei der spontanen Kondensation reiner Wasserdampfes in Lavaldüsen. *Forsch. Ing. Wes.*, Vol. 37, No. 5, 1971, pp 146-157.

4.31 Bakhtar, F. & Mohammadi Tochai, M.T.: An investigation of the Two Dimensional Flows of Nucleating and Wet Steam by the Time Marching Method. *Int. Journal Heat & Fluid Flow*, Vol. 2, No. 1, March 1980, pp 5-8.

4.32 Snoeck, J.: Calculation of mixed flows with condensation in one dimensional nozzles. In: *'Aero-Thermodynamics of Steam Turbines'*, 1981, ASME H0023, 1981, pp 11-18.

4.33 Snoeck, J.: Ecoulement à Condensation en Grille d'Aubes de Turbine. *Ph.D. Thesis, Université Catholique de Louvain*, 1982.

4.34 Marble, F.E.: Dynamics of a Gas Containing Small Solid Particles. *Combustion and Propulsion, 4th AGARD Colloquium*, 1963, pp 175-213.

4.35 Keenan, J.H.; Keyes, F.G.; Hill, P.E.; Moore, P.G.: Steam Tables. New York, *John Wiley*, 1969.

4.36 Magerfleisch, J.: Umkehrbare Gleichungen zur Berechnung von Entspannungsvorgängen in Wasserdampf. *BWK*, Heft 10, Oktober 1979, pp 403-407.

4.37 Young, J.B. & Bakhtar, F.: Non-Equilibrium Effects in Wet Steam Turbines. In: *'Steam Turbines for Large Power Outputs'*, VKI LS 1980-06.

4.38 Schmidt, E. (Editor): Properties of Water and Steam in SI Units. Berlin, *Springer Verlag*, 1969.

4.39 Hedbäck, A.J.W.: Theorie der spontanen Kondensation in Düsen und Turbinen. *ETH*, Mitt. Inst. Thermische Turbomaschinen Nr. 20, 1982.

4.40 Gyarmathy, G.: The Spherical Droplet in Gaseous Carrier Streams : Review and Synthesis. In: *'Multiphase Science and Technology'*, Vol. 1, Washington, Hemisphere Publ. Corp., 1982.

4.41 Arts, A.: Cascade Flow Calculations Using a Finite Volume Method. In: *'Numerical Methods for Flows in Turbomachinery Bladings'*, VKI LS 1982-05; also VKI Preprint 1982-12.

4.42 Sieverding, C.H.; Decuypere, R.; Hautot, G.: Investigation of Transonic Steam Turbine Tip Sections with Various Suction Side Blade Curvatures. In: *Design Conference on Steam Turbines for the 1980's*, Inst. Mech. Engrs., London, C 195/79, 1979, pp 241-252; also VKI Preprint 1979-02.

4.43 McDonald, P.W.: The computation of transonic flow through two-dimensional gas turbine cascades. *ASME Paper 71-GT-89*.

CHAPTER 5

Aerodynamic Development of Bladings

5.1 EXPERIMENTAL FACILITIES FOR BLADING DEVELOPMENT

D.H. Evans

Steam turbine blade testing can be broadly categorized into two classes, one of verification of mechanical and thermal performance of turbines and components and the other of broadening the understanding of turbine performance and expanding a technology data base. Verification testing ranges from linear cascades to full size turbines in the laboratory and field. Technology development is achieved with the same facilities but is enhanced by test facilities for either special purposes or with considerable flexibility that allows operation at conditions rarely or never seen by a conventional turbine.

One of the most fundamental test facilities is the linear cascade. A group of blades is placed in a stream of air or steam and the aerodynamic performance is determined in a variety of ways. Flow capacities range from a subsonic air cascade (Fig. 5.1.1), to a transonic air cascade. The subsonic cascade is an open loop facility that includes a blower and is a relatively inexpensive facility to operate. Typical information from this facility consists of blade losses based on traverse data in a plane just downstream of the trailing edges. A single row of blades up to six inches (0.15 m) high can be tested. This height offers sufficient flexibility to evaluate various aspect ratios and to separate profile losses from secondary flow losses at the endwalls.

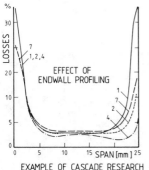

EXAMPLE OF CASCADE RESEARCH

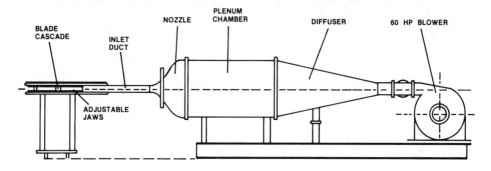

FIGURE 5.1.1 Subsonic air cascade

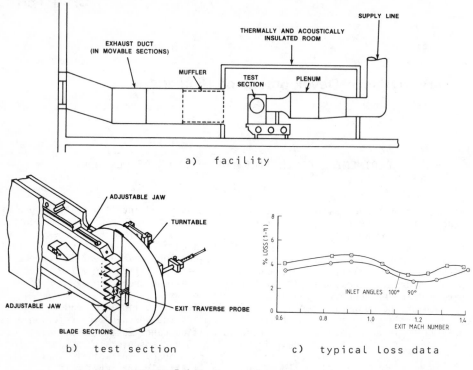

a)   facility

b)   test section   c)   typical loss data

FIGURE 5.1.2   Transonic air cascade

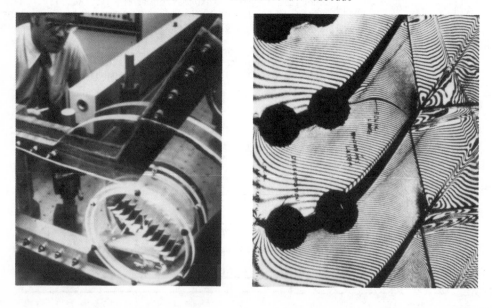

FIGURE 5.1.3   Transonic optics cascade

For higher Mach number operation the transonic cascade, shown in figure 5.1.2, is well suited for testing low pressure turbine blading. The cascade is an open loop facility supplied by an air compressor with an exhaust duct to the outside and a noise enclosure around the cascade. An optics system permits still or moving pictures of shock patterns or pressure distributions using schlieren or interferometry methods. Specially instrumented blades make possible pressure measurements at the blade surfaces. Programmed traverse mechanisms are used to automatically traverse at blade mid height for profile losses and in the endwall regions for secondary flow losses. Typical loss data are shown in figure 5.1.2-c.

Special purpose cascades such as the optics cascade shown in figure 5.1.3 offer unique advantages. It operates in a suction mode up to Mach numbers of 1.5. This type of operation is relatively quiet and at low temperature hence providing better operating conditions than the transonic cascade for delicate optical work. It is conductive to shadowgraph, schlieren or interferometry for still and moving pictures. This facility is not used for loss measurements although blade surface pressures are obtained from interferometric patterns as shown in figure 5.1.3-b.

Another special purpose cascade is the rotating water table shown in figure 5.1.4. This facility makes use of a hydraulic analogy to model two dimensional compressible flow [5.1]. It has proved valuable to obtain quick qualitative information on the effect of unsteady flow in turbine stages. The effect of stage geometry on aerodynamic excitation forces can be determined for full and partial admission. Blade forces are measured and motion pictures provide additional insight into the flow and reflected wave patterns.

FIGURE 5.1.4 Rotating water table

Operating conditions unique to steam turbines are evaluated in a steam laboratory dedicated to wet and dry steam tests in a transonic cascade and a nozzle facility. In addition to the measurement of blade losses, information is obtained on flow instability due to shock wave steam condensation-boundary layer interactions. A series of experiments have been conducted in a one-dimensional Laval nozzle to determine the nature of self-excited fluctuations due to shock wave/boundary layer interaction. Shock wave and pressure fluctuations have been recorded with frequencies and level of activity differing for wet and dry steam. Results have been observed of shock wave interaction with incipient condensation near the nozzle throat. Observations from these tests add to the understanding of excitation forces that turbine blades are exposed to when operating in a transonic flow regime with steam conditions near the saturation line. Sodium chloride deposition, which occurs very close to the steam saturation line, is studied. This information is applicable to the reliability of turbine blades that operate near the saturation lines.

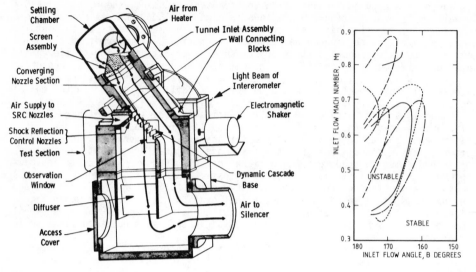

a)   blowdown tunnel and dynamic cascade   b)   instability regions

FIGURE 5.1.5 Blowdown tunnel for the study of stall flutter

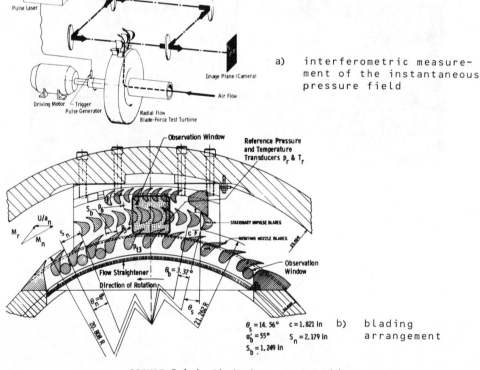

a)   interferometric measure-
     ment of the instantaneous
     pressure field

b)   blading
     arrangement

FIGURE 5.1.6 Blade-force test turbine

Shown in figure 5.1.5 is a blowdown wind tunnel used for the study of stall flutter [5.2]. A cascade of blades is mounted dynamically and connected to a magnetic shaker. Tests are run over a range of inlet Mach numbers and flow inlet angles. The facility is instrumented so that positive and negative aerodynamic damping can be determined depending on whether energy is transferred from the shaker to the blades or from the blades to the shaker. An observation window and interferometer system allows visual studies to increase understanding of this phenomenon.

Tests of a number of blade sections from different last rotating rows of low pressure turbines over a range of operating conditions have identified stable and unstable operating regimes. Figure 5.1.5-b shows test results for blades with different characteristics such as chord, camber and inlet angle. Each blade section exhibits different sensitivity to the stall flutter phenomenon. Sufficient data has been acquired to deduce empirically what blade characteristics are desired for a blade to be insensitive to or free from flutter.

Another facility dedicated to non-steady flow is a Radial Flow Air Turbine (Fig. 5.1.6), that utilizes the same blowdown system and Mach-Zehnder interferometer as the stall flutter facility. The interferometric measurement system is shown in figure 5.1.6-a. The interferometer light sources are a 150 MW ruby laser having a pulse duration of 0.02 sec., and a 15 mW He-Ne gas laser with a strobing cability up to 10 kHz. Figure 5.1.6-b shows the blading arrangement. The objective of this facility is to measure wake forces and shock forces on control stage blades during partial admission. The Radial Flow Turbine can accommodate a nozzle flow Mach number up to 1.6.

Multistage turbine test facilities range from small units of several thousand horsepower to full size low pressure turbines with shaft output to around 70,000 HP (52000 kW). The smallest unit is suitable for control stage, high pressure and intermediate pressure turbine blades. Turbine thermal and mechanical performance can be determined over a range of operating conditions. Power is absorbed by a disc water brake which permits operation over a broad spectrum of steam loads and speeds. The ability to test multiple stages provides test data for conditions similar to actual operating conditions. The facility, shown in figure 5.1.7, is instrumented for moisture trajectory studies using fibre optics for internal lighting and borescopes connected to cameras for still and high speed motion photography. The purpose of this test was to observe moisture movement across turbine blades and when being shed from blade trailing edges [5.3]. The turbine was designed with conditioning stages upstream of the test stage so that the moisture observed was formed naturally through the turbine expansion process.

An example of the type of photographic information acquired from this test is shown in figure 5.1.7-b. The view is near the trailing edge. The light is passed through a fibre optic tube and projected through the wake leaving the stationary blade. The light is aimed at a borescope which is attached to the camera. The size and position of the water droplets, which appear as black dots, are measured by an optical comparator. This data is then compared to analytical predictions of drop breakup in the wake region.

Careful test procedures were necessary to obtain this type of information in an actual turbine. The borescope mirror had to be kept clean for quality photographs. Screening gases were directed around the mirror to prevent moisture from collecting. Focusing the camera was especially difficult. After the turbine was assembled and photographic equipment installed, it was necessary to focus the camera in space on the plane defined by the direction of the blade wake. An object placed in the wake was unacceptable since it would disturb the floor. A retractable device was designed that was pushed into the focus plane by air pressure. A transparent flag with markings provided a target for focusing. Careful alignment of the flag, prior to turbine assembly, was essential to ensure that the flag was placed in the plane of the stator blade wake.

Shown in figure 5.1.8 is a scale model multistage low pressure test facility [5.4]. A water brake rated at 25,000 HP (18700 kW) and 10,000 RPM, absorbs shaft

D.H. Evans

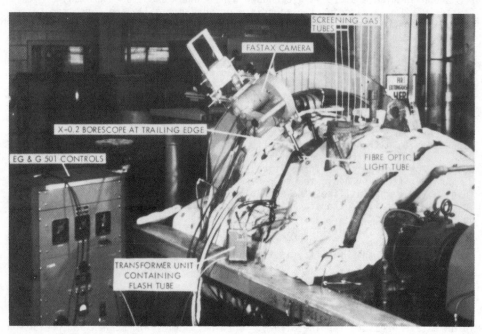

a) photographic equipment when testing

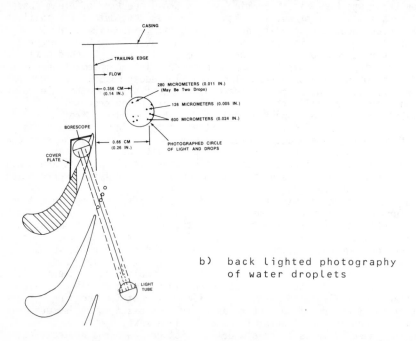

b) back lighted photography
    of water droplets

FIGURE 5.1.7 Wet vapour turbine test rig

output. The outer shell absorbs the vacuum loading making it relatively inexpensive to test components such as exhaust diffusers that can be interchanged easily. Also, access holes for traverse probes can be quickly added to suit the blading to be tested.

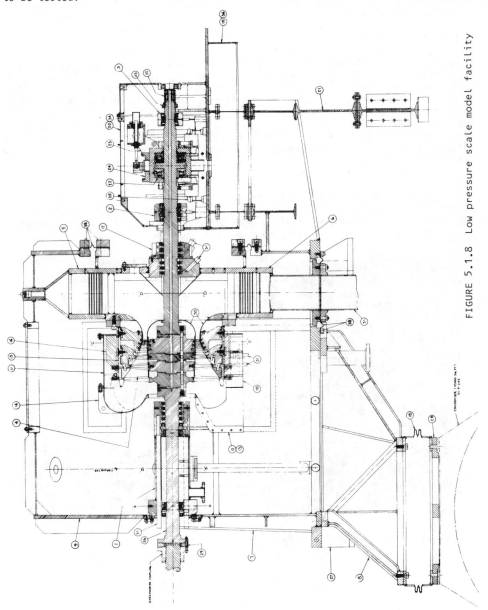

FIGURE 5.1.8 Low pressure scale model facility

Verification testing has been performed on full size low pressure turbines installed in a Low Pressure Steam Turbine Laboratory [5.5]. Due to the large capacities required the laboratory was constructed in an existing electricity generating station with sufficient boiler and condensing capabilities. The laboratory was essentially a complete power plant that included provision for extractions, feed water heaters, control systems and other auxiliary equipment normally found in a power plant. Turbine thermal performance was based on available heat drop and shaft power measured by a 68,500 HP (51100 kW) water brake. Tests have been conducted on full scale models of fossil 3600 RPM turbines with 31 inch (0.79 m) last row blades and half scale models of nuclear turbines having 44 inch (1.12 m) and 40 inch (1.02 m) last row blades.

This type of facility allows testing of full size turbines at conditions that cannot be easily produced in field units. The turbine can be operated at high power levels over a broad speed range and at abnormally high back pressures. Information from such tests enhance the understanding of blade mechanical performance. The turbine thermal performance is more accurately defined in the laboratory than in the field. The low pressure turbine performance determined by a field economy test is deduced from a full heat balance of all turbine components. Hence the accuracy of the low pressure turbine efficiency is affected by the accuracy of many other measurements of the entire turbine system. In the laboratory the shaft power measurement is a direct determination of the turbine output.

Test data from a full size turbine and a model turbine of the same geometry is useful to develop scaling laws. This is especially important in steam turbine testing where two phase flows are encountered. Moisture formation and movement do not scale according to conventional rules such as those based on Reynolds number. Therefore, unique scaling laws are developed for steam turbines when data is available from both full size and model size turbines.

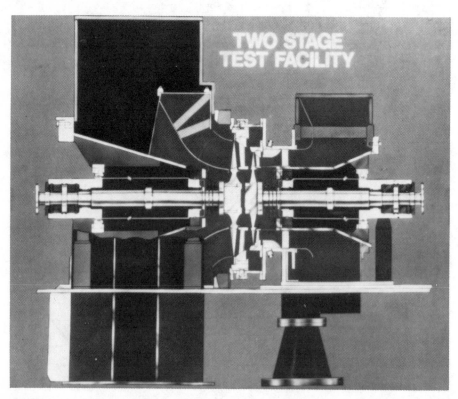

FIGURE 5.1.9 Two stage test facility

A two stage facility with considerable flexibility, shown in figure 5.1.9, has separate shafts to independently measure power output of each stage. Blades up to 18 in. (0.46 m) can be tested through a broad range of conditions. The power absorption brakes are rated at 15,800 HP (11,200 kW) and 10,000 RPM. The down-stream test stage can be driven by a steam turbine to simulate far-off design operation for the last stage of a low pressure turbine. Data for this operating regime is valuable for stall flutter analysis and verification of flow field analysis calculations. Each turbine stage has two journal bearings and one thrust bearing which are held in a movable housing that is supported in special bearings. The bearing torque is measured by a load cell so that bearing losses can be calculated for each stage. An independent check of bearing losses is made based on oil flow rate and inlet and exit oil temperature. An accurate determination of bearing losses, especially for a single stage power measurement is essential to establish the absolute level of thermal performance.

The upstream stage serves primarily as a conditioning stage to create multi-stage flow conditions into the test stage. This stage, the inlet cylinder, and dynamometer are mounted on a common movable platform that is pulled away from the test stage and exhaust cylinder. Only the main steam line joint need be broken. For low pressure, low temperature tests only six bolts are required between inlet and test section. This quick access feature allows rapid repair, modification, or addition of instrumentation and relatively fast change-over of the test blading. Access holes are provided for traverse probes and two viewing windows permit visual observations during operation.

Approximately 600 measurements can be made of pressure, temperature, flow, speed and torque and 144 signals from rotating strain gages and fast response pressure transducers that are transmitted through slip rings mounted on each end of the test turbine. The data acquisition system reflects the latest technology in test measurement systems. This new system has a number of features which make it far superior to earlier systems. For example, the system can sample up to 10,000 points of data per second as well as print out individual test points every 15 seconds. Stage efficiency is calculated every five minutes based on accumulated data over that time period. The sophistication of the data reduction calculations is comparable to what previously was reserved for computation on a main frame computer. In addition to reducing the data, the system is also capable of calculating the predicted stage performance for the specific test conditions. Therefore, the test engineer can make an accurate and quick determination of how test and prediction compare without relying on generalized prediction curves. The system has sufficient storage capacity to hold the experimental data from several model tests.

The data from this facility makes possible accurate analysis of turbine and blading performance. Independent measurement of stage power output maximizes the precision of measured efficiency thus affording the opportunity to assess the effect of modest design changes.

For every new design of a tuned low pressure turbine blade, verification testing of the natural frequencies is performed in a vacuum facility with a steam turbine drive 5.6 . A tuned blade is one that is designed such that the natural frequencies of the lower modes of vibration do not coincide with harmonics of turbine design speed.

The blades are operated over a wide range of speeds and are excited by steam jets. A single jet is used for the lower modes and multiple jets for higher modes that may be excited by upstream stator blade wakes. Decay rates of vibration amplitude are measured to verify damping properties. A Cambell diagram is developed that determines what the natural frequencies are for the blade at design speed. Sufficient data is taken to establish statistically the frequency limits that all manufactured blades must fall within.

Verification tests are periodically performed on field units for mechanical and thermal performance. Over the road trailers are instrumented to perform full ASME Economy Tests and telemetry tests where strain gage signals are sent from ratio transmitters mounted in turbine balance holes. One trailer is equipped

with all the hardware required for installing strain gages, transmitter and all
the associated equipment for running a field telemetry test. The trailers are
placed near the turbine side by side to form a room 16 ft. (4.9 m) by 32 ft.
(9.8 m).

The ultimate determination of the thermal performance of a large central station turbine-generator unit is the field unit economy test. Conventional performance testing of a large steam turbine requires the recording of data by up to 50
individuals stationed throughout the power plant. Despite instruction in the use
of test instrumentation and good data taking procedures and also close test super-
vision, errors occur. The data recorded is in raw form such as pressure in milli-
meters of mercury. Unit conversions and instrument calibration have to be accounted for prior to the reduction of test data into turbine heat rate and component
efficiency. Accurate calculations cannot be performed until after the test program is completed. Only limited checks are made during testing to assess the
validity of the measured data.

A specially instrumented over the road trailer is dedicated to heat rate
economy tests [5.7]. A Hewlett Packard 1000 computer is used for data acquisition
and reduction. Up to 400 data points are collected and processed by the computer.
In addition to the heat rate calculations performed by the HP 1000 computer, test
data is relayed to a main frame computer for further analysis. Even during the
data collection process the engineers can request the computer to calculate the
turbine cycle performance without interrupting the collecting process. The trailer
also serves as a test control center. Portable radio transmitters and the power
station intercom system are used to communicate with station personnel to correct
any problems with instrumentation and equipment.

This data acquisition system has been applied to the heat rate tests of a
nuclear unit. Some data takers were used to record data in the conventional
method for comparison to that recorded by the data acquisition system. When discrepancies were noted, invariably the error was found in the manual data recording method.

In addition to the standard heat rate tests, special tests were run at the
customer's request to determine the effects of several cycle components on performance. The results of changing a cycle parameter were immediately known with
the data acquisition system.

The use of a mobile test facility provides test engineers with increased data
visibility and better test manageability. The test supervisor can make better
informed judgements as to the progress and quality of the test. The amount of
manual data required is significantly reduced with a corresponding reduction in
cost.

Many other test facilities are used to evaluate turbine components such as
exhaust hoods, inlets, bearings, valves and controls. Much of the development
process of the high performance steam turbine has been dependent on a large data
base of test information gathered from laboratory facilities and field units.

5.2 AERODYNAMIC CHARACTERISTICS OF LAST STAGE BLADE PROFILES

C.H. Sieverding

5.2.1 Last Stage Flow Conditions

As described in chapter 1, the final stage moving blade of large steam turbines
presents special problems of aerodynamic design. These problems arise mainly from
the use of hub/tip diameter ratios as low as 0.5 in order to provide maximum flow
annulus area at minimum tip diameter. Consequently flow velocities and angles,
and hence blading profiles, vary considerably from root to tip of the final stage.
This chapter considers the aerodynamic characteristics of the blade profiles used.
As the flow field is complex, it is customary, at present, when developing blade

profiles to assume the flow behaves as a single-phase medium; the additional effects of condensation are considered in chapter 4.

The range of flow conditions in last stage bladings of turbines of large power output are best illustrated by two examples taken from the literature.

Example 1 : A KWU stage with a 950 mm long blade with a tip diameter of D_T = 3500 mm and a mean diameter to length ratio D_M/H = 2.7, running at 3000 RPM [5.8]. The angle and velocity distributions are shown in figure 5.2.1. The main flow characteristics are :
- constant α_2 over 60% of the stator exit blade height;
- constant M_3 over 50% of the rotor exit blade height;
- maximum absolute rotor outlet Mach number $M_3 \approx 0.6$;
- relative rotor inlet Mach number $M_{W2} \leqslant 0.8$;
- maximum absolute exit velocity from stator $M_{2,H}$ = 1.33;
- maximum relative exit velocity from rotor $M_{W3,T}$ = 1.53;

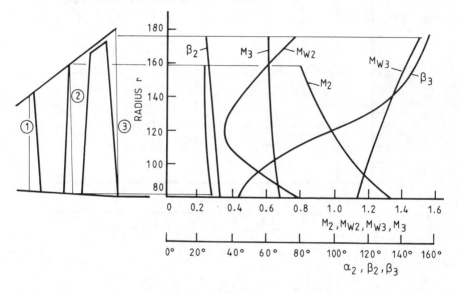

FIGURE 5.2.1   Flow conditions for a KWU 950 mm blade design at 3000 RPM [5.8]

Example 2 : A BBC stage with a 1200 mm blade with D_T = 4452 mm and D_M/H = 2.71, running at 3000 RPM [5.9]. The blade sections and spanwise angle and velocity distributions are shown in figure 5.2.2. Particular features are : (a) the high relative rotor inlet velocities at hub (M_{W2} = 1.03) and tip, (M_{W2} = 1.13) and (b) a constant relative rotor inlet angle $\beta_2 \approx 70°$, extending from the hub to 20% of the blade height. The corresponding approximate velocity triangles at hub and tip, derived by Foster [5.10], are shown in figure 5.2.2c. The blade root flow conditions imply presumably a considerable stream tube divergence between blade inlet and outlet.

The two examples give a fair idea of the aerodynamic problems involved in the development of typical last stage bladings. The aerodynamic design has to be accomplished within the narrow limits set by mechanical requirements. The optimization of the rotor blade is particularly arduous, the high centrifugal stresses imposing a very strong variation of the blade sectional area from hub to tip. This aspect is illustrated in table 5.2.1 which shows typical values of chord lengths and pitch to chord ratios at rotor hub and tip as used by various manufacturers. The values are evaluated from photographs and figures in the published literature and are therefore approximate.

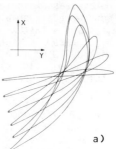

a)   Rotor blade sections

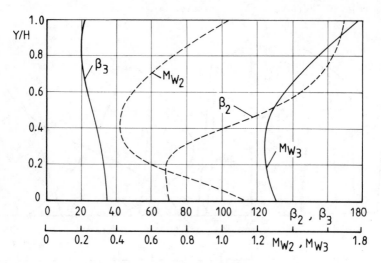

b)   Flow conditions in rotor

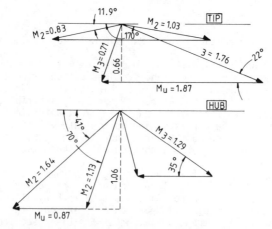

c)   Velocity triangles derived by Foster [5.10]

FIGURE 5.2.2   Flow conditions and blade section design for 1200 mm
                BBC blade at 3000 RPM [5.9]

H (mm)	c_T (mm)	c_H (mm)	$(g/c)_T$	$(g/c)_H$	RPM	Firm
910	~ 130	190→230	0.9	0.25→0.3	3000	EE[†]
1000	122→130	~ 280	~ 1.0	~ 0.21	3000	CEM[††]
900	~ 145	~ 280	~ 0.95	0.23	3600	TOSHIBA
1200	155	~ 300	1→1.1	0.24→0.26	3000	BBC

[†] : English Electric, at present GEC
[††] : Compagnie Electro-Mécanique, at present Alsthom

TABLE 5.2.1 Geometric characteristics of last stage rotor blade rows

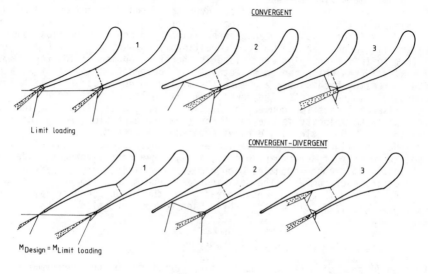

a) Shock patterns at design and off-design

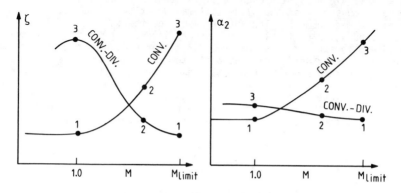

b) Basic performance curves

FIGURE 5.2.3 Basic shock patterns and performance curves for
convergent and convergent-divergent blades

5.2.2 Transonic Turbine Cascade Performances

Depending on the outlet Mach number range the blades are designed with convergent or convergent-divergent blade passages. Convergent bladings are used typically up to outlet Mach numbers $M_2 \simeq 1.3$ to 1.4 while convergent-divergent blade passages are preferred for higher Mach numbers. Figure 5.2.3a presents schematicall the shock configurations for both types of blades at various off-design condition The nominal operating point for the convergent-divergent blade is chosen to be equal to the limit loading point. The nominal operating point for the convergent blade is supposed to be around $M_2 = 1.0$.
        The qualitative angle and loss variations are sketched in figure 5.2.3b.
The total profile losses are the sum of :
(a) shock losses inside the blade passage;
(b) boundary losses along pressure side and suction side;
(c) trailing edge losses; dependent on trailing edge thickness and base pressure;
(d) downstream mixing losses.
The last loss component includes both downstream shock losses and losses due to the viscous mixing process. The combined action of shock waves, expansion waves and viscous forces cause a gradual transition from non uniform flow conditions in the trailing edge plane to uniform conditions far downstream. The degree of non uniformity in the trailing edge plane is therefore an indication for the amount of mixing losses which can be expected. For convergent cascades it will increase with increasing Mach number and make the mixing losses progressively the most important loss component. The downstream mixing losses are responsible for the rapid rise of losses for convergent bladings for $M_2 > 1$. Since the increase of the downstream velocity is due to a Prandtl-Meyer expansion around the blade trailing edge the outlet angle rises continuously with $M_2 > 1$.
        The evolution of the profile losses for convergent-divergent blades is opposite to that of convergent blades, i.e., convergent-divergent blades have high losses at transonic outlet Mach numbers and minimum losses at the supersonic design Mach number. The loss increase from M_{design} to $M_2 = 1$ is mainly due to shock induced thickening and separation of the blade boundary layer. The outlet flow angle is little affected by a Mach number variation.

5.2.2.1 Effect of Rear Suction Side
           Curvature in Convergent Cascades

The significance of the rear suction side curvature for the performance of convergent cascades was recognized since the early 1950's. Based on some unpublishe work at NGTE Ainley & Mathieson [5.11] stated that the profile loss for blades with large curvature of the blade tail may be at $M_2 = 1$ as much as four times the loss at low Mach number. The authors ascribed the increase of losses to shock induced separation of the laminar boundary layer. Blades with a curved tail are more likely to exhibit a laminar boundary layer since the curvature tends to move the point of peak suction side velocity towards the trailing edge in comparison to straight-backed blades. Hence blades with rear suction surface curvature are likely to have higher losses when the surface Mach number exceeds critical velocity. This interpretation of the test results was confirmed by the fact that an artificial transition of the boundary layer on the front part of the suction side brought the losses back to the low Mach number level. The influence of outlet Mach number and rear suction side curvature was presented by Ainley & Mathieson under the form of a non dimensionalized loss coefficient Y/Y_{ref}, where Y_{ref} prese the loss at $M_2 = 0.5$ (Fig. 5.2.4). This presentation hides the fact that straigh backed blades exhibit in general higher losses at low Mach numbers than blades with a curved tail.
        Craig & Cox [5.12] published their performance prediction method exactly 20 years after Ainley & Mathieson. With the analysis of steam turbine stages in view, the authors extended their correlation to high supersonic Mach numbers.

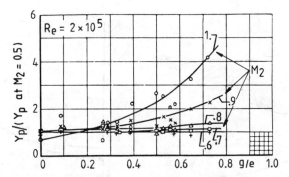

FIGURE 5.2.4   Effect of rear suction side curvature on profile losses
due to Ainley & Mathieson [5.11]

For $M_2 > 1$ a fundamental additive correction is made to the basic subsonic loss
coefficient. The correlation of this correction term, referred specifically to
convergent profiles designed with a straight rear suction side, is presented in
figure 5.2.5. These losses are due to the homogenization process between the non
uniform flow conditions in the trailing edge plane and the uniform flow conditions
existing far downstream. For profiles with a curved tail the correlation foresees
a further additive loss over and above that given for straight-backed blades
(Fig. 5.2.5b). This second correction term becomes effective for $M_2 > 0.8$.

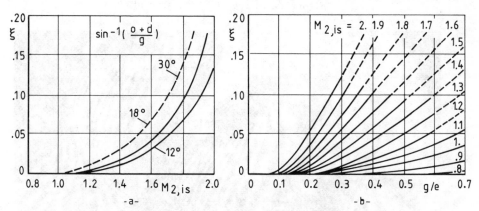

FIGURE 5.2.5   Effect of outlet Mach number and rear suction side
curvature on profile losses, due to Craig & Cox [5.12]

The curvature of the rear suction side affects the profile losses in
distinctly different ways. In addition to the higher probability of a shock
induced laminar boundary layer separation mentioned by Ainley, a curved back
increases the maximum suction side velocity compared to that of a straight backed
blade. This is particularly clear for the maximum loading conditions as illustrated in figure 5.2.6. The degree of non uniformity at the trailing edge increases and so do the losses associated with the downstream homogenization process. A simple indication of the degree of non uniformity at the trailing edge
plane is the pressure difference across the trailing edge. For very large pressure differences the suction side flow may separate from the blade surface before
reaching the trailing edge as shown in the interferogram in figure 5.2.7 from
[5.13]. In most applications, convergent cascades do not operate at limit loading

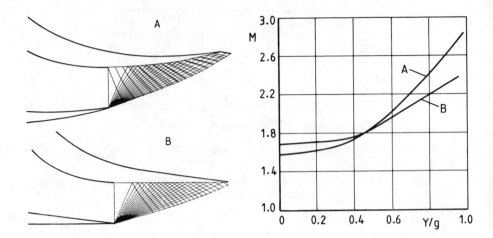

a) Wave diagrams for curved and   b) Velocity distribution in
    straight rear suction side   TE plane at limit loading

FIGURE 5.2.6 A straight backed suction surface contributes to limit the
acceleration of the rear suction flow and to reduce the degree of non-
uniformity in the trailing edge plane

FIGURE 5.2.7 Flow separation near the trailing edge due to
overacceleration of flow along convex suction surface, from
Nagayama et al., Mitsubishi [5.13]

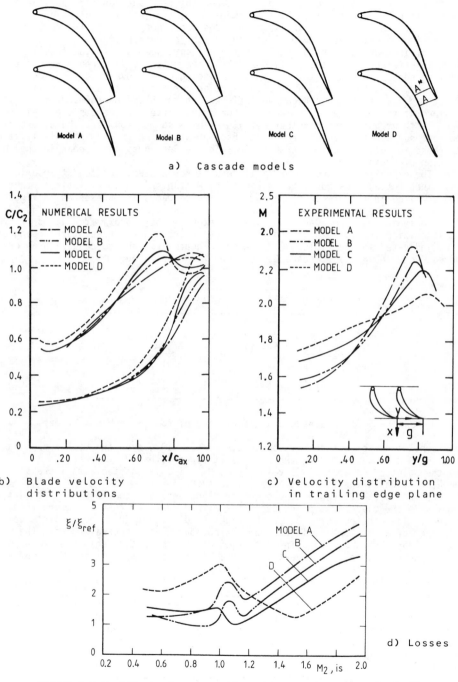

a) Cascade models

b) Blade velocity distributions

c) Velocity distribution in trailing edge plane

d) Losses

FIGURE 5.2.8 Effect of rear suction side curvature on cascade flow characteristics, Nagayama et al., Mitsubishi [5.13]

conditions but a lower pressure ratios. Then the rear suction side pressure distribution is characterized by a pressure jump, the intensity of which is a function of the strength of the trailing edge shock impinging on the suction side. An experienced designer may evaluate the blade performance from the blade pressure distributions by considering both the pressure jump due to the trailing edge shock interference with the suction side and the pressure difference across the trailing edge. The strength of a shock impinging on a convex suction side is reduced compared to that of a straight suction surface but this local positive effect may have an overall negative influence on the blade performance if it is obtained at the expense of an increased pressure difference across the trailing edge. Without going into further detail let us also mention here the effect of the rear suction side curvature on the base pressure, which is particularly strong at transonic outlet Mach numbers [5.14]. This effect does not appear in the correlation of Craig & Cox.

The experimental and theoretical results for a series of four transonic steam turbine cascades published recently by Mitsubishi [5.13] illustrate the issues raised in the preceding paragraph. The cascades are presented in figures 5.2.8. The blade channels of cascades A, B and C are convergent. The blades differ only in their rear suction side curvature. The approximate values of g/e, characterizing the curvature, and the corresponding blade surface turning angles ε are indicated in the table 5.2.2. A convergent-divergent cascade with an area increase of $A/A^{\star} = 1.1$ (cascade D) is shown for comparison.

Blade	A	B	C	D	
g/e	$\sim 0.12^{1}$	$\sim 0.24^{1}$	0	-	^{1} estimated
$\varepsilon[^{\circ}]$	13°	6°	0	-	
divergence $A/A^{\star}$	1	1	1	1.1	

TABLE 5.2.2 Cascade characteristics of four Mitsubishi blades

The Reynolds number was nearly constant, $Re = 5 \times 10^{5}$, over the Mach number range, $0.8 \leqslant M_2 \leqslant 2.0$. Summarizing the results it appears that
(a) the point of maximum blade velocity at $M_2 \approx 1.0$ is shifted with increasing curvature towards the trailing edge (Fig. 5.2.8b);
(b) the pressure difference across the trailing edge and the degree of non uniformity in the blade exit plane decreases with decreasing rear curvature (Fig. 5.2.8c). The Mach number drop on the right-hand side of the figure points to the existence of a shock which is caused by a boundary layer separation near the trailing edge (Fig. 5.2.7). The extension of the separated region decreases with decreasing curvature. The overall tendencies are confirmed at high Mach numbers by the results of the convergent-divergent blade D;
(c) blade C with straight rear suction side has higher losses at subsonic Mach numbers than blades B and A but lower losses for $M_2 > 1$ (Fig. 5.2.8d). The local loss peaks around $M_{2,is} \approx 1.1$ for blades A and B are probably caused by a separation of the laminar boundary layer near the trailing edge under the influence of normal shocks standing on the rear blade surface. The separation disappears when these shocks move to the trailing edge. Cascade D exhibits a typical performance curve for convergent-divergent bladings. Lowest losses occur close to the exit Mach number $M_2 = f(A/A^{\star})$.

Occasionally use is made of convergent cascades with convex-concave rear suction sides. The wave system for this design is shown in figure 5.2.9, from Dejc & Trojanovskij [5.15]. Compared to a straight-back blade the point of minimum losses is shifted to higher Mach numbers. The authors recommend the use of this type of profile for exit Mach numbers M_2 = 1.3 to 1.4. Operated at transonic outlet Mach numbers, these blades exhibit a similar loss peak as convergent-divergent bladings.

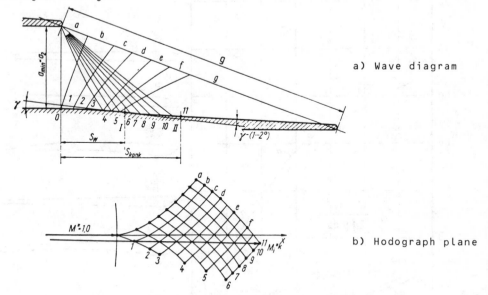

a) Wave diagram

b) Hodograph plane

FIGURE 5.2.9 Convergent cascade with convex-concave rear suction side from Dejc & Trojanovskij [5.15]

5.2.2.2 *Performances of Last Stage Rotor Blade Sections*

Tip sections

The development of special transonic blade profiles for the last stage of large steam turbines started in the early 1960's. In 1967, Meyer of BBC [5.16] published the results of an extensive investigation of a large number of blade profiles for the tip section of a last stage rotor (Fig. 5.2.10). The blades were designed for α_1 = 155°, M_1 = 0.63, α_2 = 29° and M_2 = 1.71. The data were obtained from measurements on a rotating cascade. The most striking result of this study was the close relation between blade efficiency and degree of overlapping between two successive blades.

In view of the high supersonic outlet Mach numbers, tip sections should be designed with a convergent-divergent internal flow passage. The tip profile in figure 5.2.11a designed for M_2 = 1.7, shows fairly well the diffulties inherent in such a design (taken from a schlieren photograph in [5.17], the dotted part was added by the author). The corresponding blade velocity distributions for dry inlet steam conditions are presented in figure 5.2.11b. The small pressure difference across the trailing edge, the moderate recompression after the trailing edge shock impingement on the suction side and a maximum suction side velocity, which is little in excess of the downstream velocity, are indications of a high efficiency near design conditions. The corresponding pressure distributions for wet inlet conditions are shown in figure 5.2.11c. It is noticeable that the wet steam conditions appear to reduce the shock strength. This wet steam effect is confirmed in chapter 4.2 by Snoeck, who uses a time marching finite volume technique to calculate condensing flows through transonic rotor tip sections.

Figure 5.2.12 summarizes the results of an experimental investigation of three tip sections with various suction side blade curvature [5.18]. The main

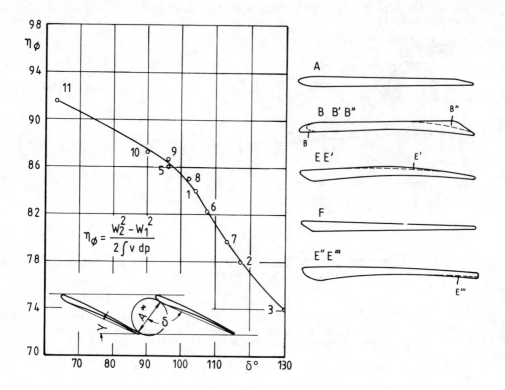

$$\eta_\phi = \frac{W_2^2 - W_1^2}{2 \int v\, dp}$$

SECTION	TEST N°	CASCADE	PITCH/CHORD	δ (DEGREE)	γ (DEGREE)
A	1	1A	1.119	103.5	21
	2	2A	1.326	117	21
	3	3A	1.629	130	21
	4	4A	1.326	125	16
	5	5A	1.053	96	23
B	6	6B	1.119	107.5	25
B'	7	6B'	1.174	113	25
B"	8	6B"	1.174	101	25
B"	9	5B"	1.105	96	25
E	10	7E	1.000	90	23
E	11	8E	0.835	63	23

FIGURE 5.2.10 Influence of cascade angle δ on efficiency of rotor
blade tip sections, Meyer, BBC [5.16]

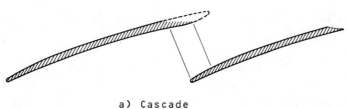

a) Cascade

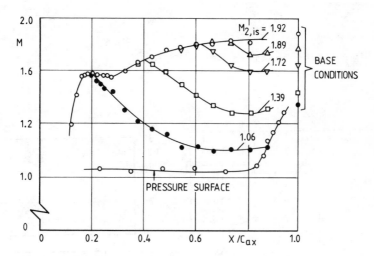

b) Dry steam condition

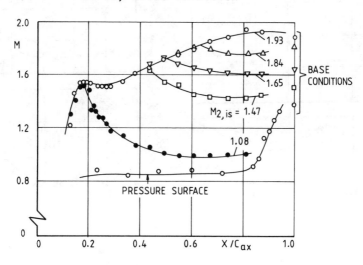

c)   Wet steam condition (6% at inlet)

FIGURE 5.2.11   Convergent-divergent tip section by Cox, GEC [5.17]

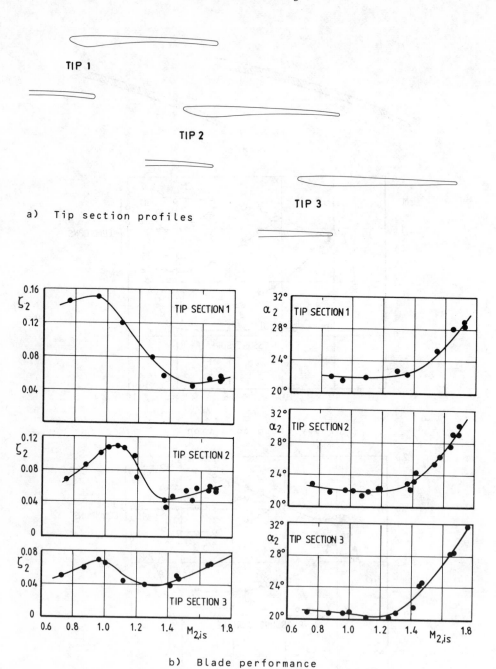

a)   Tip section profiles

b)   Blade performance

FIGURE 5.2.12   Influence of suction side curvature on performance of three
            steam turbine rotor tip sections, Sieverding et al. [5.18]

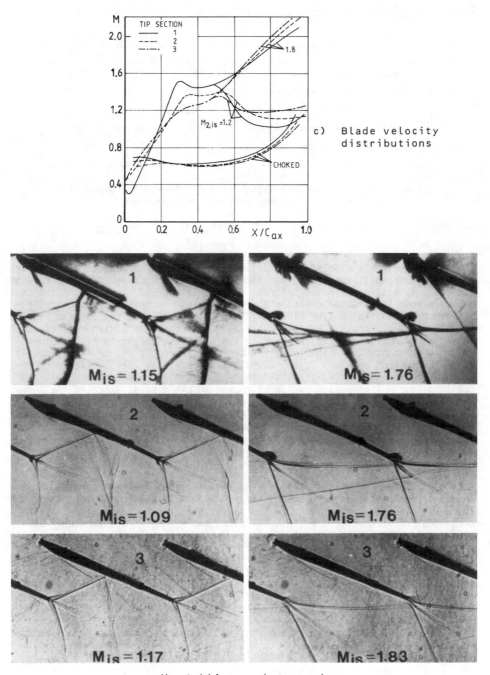

c) Blade velocity
    distributions

d) Schlieren photographs

FIGURE 5.2.12 Influence of suction side curvature on performance
of three steam turbine rotor tip sections, Sieverding et al. [5.18] (cont'd)

characteristics are summarized in table 5.2.3 :

TIP SECTION	1	2	3
stagger angle, γ	156°	156°	156°
pitch to chord ratio, g/c	0.9	0.86	0.85
throat to pitch ratio, o/g	0.333	0.333	0.333
trailing edge thickness, d/c	0.03	0.018	0.018
trailing edge wedge angle, δ	1°	2°	3°

TABLE 5.2.3 Cascade characteristics of three rotor blade tip
sections with various suction side curvature

The blades are designed for : $\alpha_1 = 156°$; $M_1 = 0.6$,

$$\alpha_2 = 29°; \quad M_2 = 1.7$$

Similar to the profile in figure 5.2.11a, <u>blade TIP1</u> is characterized by a strong convex-concave suction surface, but the internal blade passage divergence is small and there is little expansion prior to the trailing edge. The sonic point on the pressure side is situated close to the trailing edge. This blade suffers from an overexpansion on the suction side immediately downstream of the throat (Fig. 5.2.12c). Nevertheless, the losses are very moderate near design conditions, being $\sim$ 5 to 6% (Fig. 5.2.12b). For $M_2 < 1.4$ the losses begin to rise rapidly. The main reason for the high losses in the low supersonic and transonic domains are shock induced boundary layer (turbulent) separations, see schlieren photograph 5.2.12d. <u>Blade TIP3</u> has a straight suction side between the throat and trailing edge. This design avoids of course an overexpansion behind the throat but the non uniformity in the trailing edge plane increases at high outlet Mach number compared to blade TIP1. This explains also the higher losses. On the contrary the losses are much smaller for $M_2 < 1.45$. Minimum losses occur at M_2 = 1.3. <u>Blade TIP2</u> takes an intermediate position between TIP1 and TIP3.

Hub sections

Flow conditions at the root of last stage rotor blades are characterized by a low degree of reaction with turning angles of 90° to 120° and high inlet Mach numbers. The aerodynamic design is a compromise due to the limiting conditions imposed by mechanical requirements, i.e., a small pitch and a large cross sectional area governed by the centrifugal load. Hence, root sections have long passages with little or no contraction.

A first example of such blade sections is given in figure 5.2.13 published by Forster in 1965 [5.19]. It is the root section of a 686 mm (27") long blade of English Electric. The blade has fairly thick leading and trailing edges and a rather high pitch-chord ratio, g/c = 0.43. The constant Mach number lines indicate that the sonic line is quite remote from the geometric throat. It curves from near the outlet on the concave surface to well before the halfway point on the suction side. The rear suction side flow separates from the blade surface. The losses are rather high, ζ = 8 to 10% (Fig. 5.2.13b), and the efflux angle deviates considerably from the gauging angle α_2 = arc sin o/g = 45°.

A second root section designed and tested at VKI is shown in figure 5.2.14. The main characteristics are :

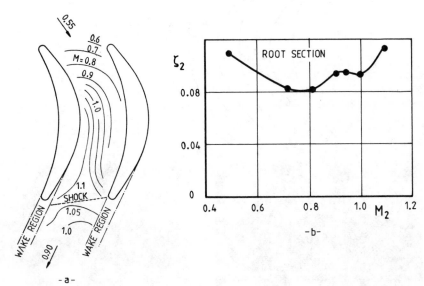

FIGURE 5.2.13 Last stage rotor root section, Forster [5.19]

pitch to chord ratio, g/c	0.32
throat to pitch ratio, o/g	0.51
internal area contraction, A/A^*	1.2
design inlet flow angle, α_1,d	45°

The blade was tested at an inlet flow angle of $\alpha_1 = 41°$. The starting point for the sonic line on the suction side is well ahead of the geometric throat; see blade velocity distributions and schlieren photographs. The losses are fairly high with a loss peak of $\sim$ 12% at $M_2 = 0.85$ but decrease significantly for higher Mach numbers; 7% at $M_2 = 1.05$. The performance could certainly be improved by eliminating the suction side velocity peak at $X/C = 0.2$.

In various cases it is difficult to avoid convergent-divergent blade channels where the position of the geometric throat is about halfway between leading and trailing edges. Profile A in figure 5.2.15, designed for a 1200 mm long blade is an example for this type of hub section (from Rzehnikov & Boitsova [5.20]. The design inlet flow angle is $\alpha_1 = 30°$, the pitch to chord ratio is $g/c = 0.305$ and the profile contour is constructed entirely with circular arcs. A second profile, B, designed for the same blade shows some important differences. By thickening the trailing edge it was possible to move the geometric throat near to the exit of the channel and also to reduce the rear suction side curvature. The variation of channel width to pitch in figure 5.2.15b indicates an increase of the quantity $(a/g)_{min}$., i.e., the maximum mass flow, by 7.5% for blade B. The redesigned channel accelerates the flow very smoothly all along the channel (Fig. 5.2.15c). In spite of the very thick trailing edge blade B performs much better than blade A (Fig. 5.2.15d). The losses due to the thickened trailing edge of blade B are apparently smaller than the losses caused by strong shock induced boundary layer separations for profile A. A decrease of the pitch-chord ratio from the design value $g/c = 0.305$ to $g/c = 0.253$ increases the losses for blade B by 50% [5.20].

Meridional flow calculations may indicate considerable divergence of the streamlines and cascade tests should be carried out under similar conditions. Forster [5.21] published some data on a hub section with both parallel and flared

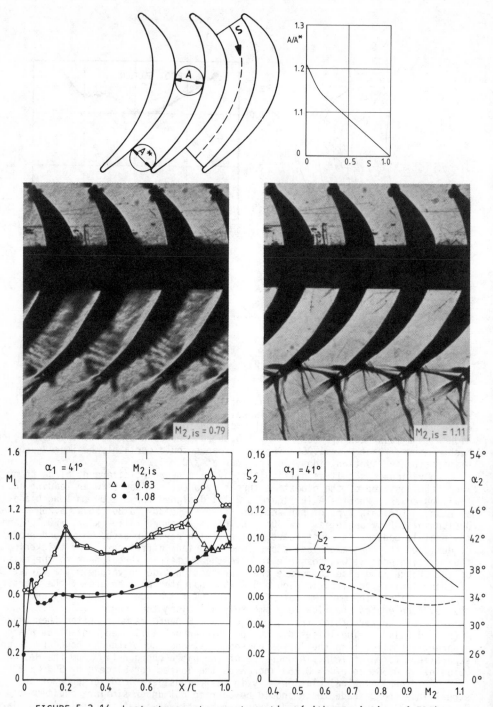

FIGURE 5.2.14 Last stage rotor root section (with permission of EDF)

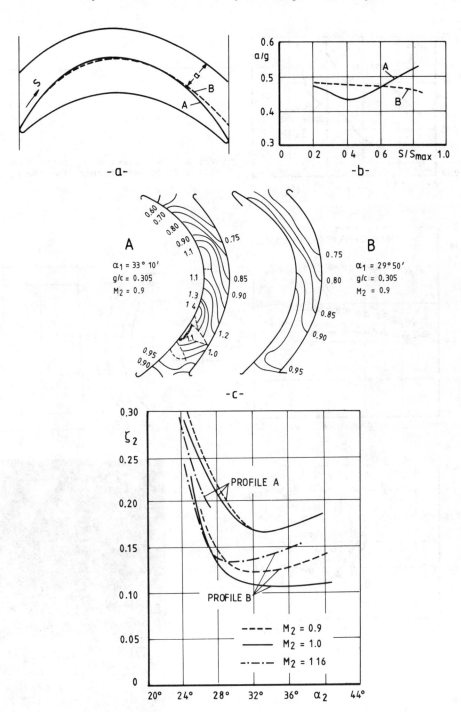

FIGURE 5.2.15 Root section design with thick trailing edge
from Rzheznikov & Boitsova [5.20]

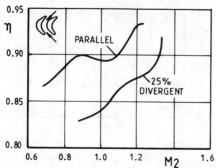

FIGURE 5.2.16 Effect of divergence on root section performance, Forster [5.21]

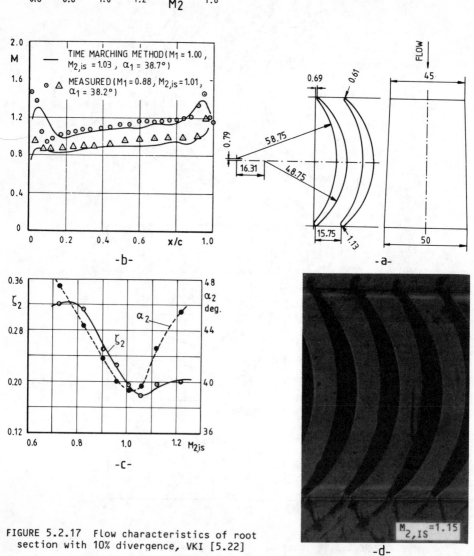

FIGURE 5.2.17 Flow characteristics of root section with 10% divergence, VKI [5.22]

side walls. In the latter case the area divergence was 25%. Figure 5.2.16 shows the efficiencies at -1.5° incidence angle. With flaring the efficiency drops by 5 to 7 points. The peak efficiency of the flared cascade is tending towards the value for the divergent area ratio of approximately M = 1.6.

The poor performance of hub section cascades with endwall flaring is confirmed by Nagao & Sieverding [5.22]. The authors tested a hub section with flared endwalls (Fig. 5.2.17) for the following design flow conditions :

$M_1 = 1.0;$ $\alpha_1 = 39.5°$

$M_2 = 1.03;$ $\alpha_2 = 41°$

The increase of the blade height by 10% between leading edge and trailing edge planes was counter-balanced by a contraction of the blade passage of the same amount. The measured and calculated (quasi three dimensional time marching finite volume method) velocity distributions show overall a fair agreement (Fig. 5.2.17b), except for an underestimation of the leading edge velocity peak by the theoretical prediction. Further information on the channel flow is given in figure 5.2.17d. The schlieren photograph confirms that except for the entrance region the channel remains shock free. The losses are nevertheless very high (Fig. 5.2.17c). Minimum losses of 18% occur at $M_{2,is} = 1.05$. The outlet angle shows a very strong dependence on the outlet velocity.

NOMENCLATURE

a	speed of sound or blade passage width
A	area
$A^{\star}$	throat area
c	chord
C	velocity
d	trailing edge thickness
D	diameter
e	rear suction surface radius
g	pitch
H	blade height
M	Mach number or
	absolute Mach number
M_W	relative Mach number
o	throat
P	pressure
Re	Reynolds number
r	radius
s	distance
U	peripheral speed
v	specific volume
Y	total pressure loss coefficient
α	absolute flow angle
β	relative flow angle

γ   stagger angle except if stated otherwise

δ   trailing edge wedge angle or cascade angle as defined in figure 5.2.10

ε   rear suction surface turning angle

ζ   loss coefficient $(C_2^2,\text{is}-C_2^2)$ / C_2^2,is

η   efficiency

ξ   loss coefficient $(C_2^2,\text{is}-C_2^2)$ / C_2^2

Subscripts

ax   axial

d   design

H   hub

is   isentropic

M   mean

p   profile

T   tip

1,2,3 in stage
      - inlet stator
      - outlet stator/inlet rotor
      - outlet rotor

1,2   in cascade
      - inlet
      - outlet

REFERENCES

5.1   Anniversary Volume for M.D. Riabouchinsky. *Publ. Scientifiques et Tech. du Ministère de l'Air*, Paris, France, 1954, pp 91-111.
5.2   Kovats, Z.: Dynamic Cascade Facility and Methods for Investigating Flow Excited Vibration and Aerodynamic Damping of Model Low Pressure Blade Groups. *ASME Paper* 79-WA/GT-6, 1979.
5.3   Evans, D.H. & Pouchot, W.D.: Flow Studies in a Wet Steam Turbine. *NASA CR* 134683, August, 1974.
5.4   Meyer, C.A.; Seglem, C.E.; Wagner, J.T.: A Turbine Testing Facility. *Mechanical Engineering*, Vol. 82, No. 7, July 1960.
5.5   Steltz, W.G.; Rosard, D.D.; Maedel, P.H.; Bannister, R.L.: Large Scale Testing for Improved Turbine Reliability. *1977 American Power Conference*.
5.6   Le Breton, A.F.: Spindle Test Raises Turbine Reliability. *Power*, August, 1960.
5.7   Southall, L.R. & Kapur, A.: Experience with a Computer Controlled Data Acquisition System for Field Performance Testing of Steam Turbines. *ASME Paper* 79-WA/PTC-1, 1979.
5.8   Maghon, H.: Die Entwicklung der Endschaufeln grosser Dampfturbinen. Paper presented at the International Meeting on *"Modern Electric Power Stations"*, Liège, Belgium, 1970.
5.9   Roeder, A.: Les Aubes Terminales du Plus Grand Corps de Turbine BP pour 3000 tr/min. *Revue Brown Boveri*, Tome 63, No. 2, Février 1976.
5.10 Forster, V.T. et al.: Development of Experimental Turbine Facilities for Testing Scaled Models in Air and Freon. *Institution of Mechanical Engineers*, Paper C53/73.
5.11 Ainley, D.G. & Mathieson, G.C.R.: An Examination of the Flow in Blade Rows of Axial Flow Turbines. *ARC R&M* 2891, 1955.
5.12 Craig, H.R.M. & Cox, H.J.A.: Performance Estimation of Axial Flow Turbines. *Proceeedings Institution of Mechanical Engineers*, 1970-71, Vol. 185, pp 32/71.
5.13 Nagayama, Kuramoto, Imaizumi: A Study of the Performance of 2D Transonic Turbine Cascades. *Mitsubishi Review*, Vol. 19, No. 2, 1982-83.
5.14 Sieverding, C.H.; Stanislas, M.; Snoeck, J.: The base pressure problem in transonic turbine cascades. *ASME Trans.*, *Series A - Journal of Engineering for Power*, Vol. 112, No. 3, July 1980, pp 711-718; also *von Karman Institute Preprint* 1979-1.
5.15 Dejc, M.E. & Trojanovskij, B.M.: Untersuchung und Berechnung axialer Turbinenstufen. Berlin, *VEB Verlag Technik*, 1973.

5.16 Meyer, H.: Transonic Flow in the Last Rotor Blade Row of a Low Pressure Steam Turbine. *Proceedings Institution of Mechanical Engineers*, 1970.

5.17 Cox, H.J.A.: Experimental Development of Bladings for Large Steam Turbines. In: *"Steam Turbines for Large Power Outputs"*, von Karman Institute LS 1980-06, April 1980.

5.18 Sieverding, C.H.; Decuypere, R.; Hautot, G.: Investigation of Transonic Steam Turbine Tip Sections with Various Suction Side Blade Curvatures. *Proceedings Institution of Mechanical Engineers*, 1979 C195/79; also *von Karman Institute Preprint* 1979-02.

5.19 Forster, V.T.: Turbine Blade Development Using a Transonic Variable-Density Cascade Wind Tunnel. *Proceedings Institution of Mechanical Engineers*, 1964-65, Vol. 179, Pt 1, pp 155-176.

5.20 Rzheznikov, Yu.V. & Boitsova, E.A.: Profiles of Long Moving Blade Root Sections. *Teploenergetika*, Vol. 15, No. 11, 1968, pp 43-46.

5.21 Forster, V.T.: Measurements in Transonic Steam Turbine Cascades. In: *"Measuring Techniques in Transonic and Supersonic Cascades and Turbomachines"*, Proceedings of the Symposium held in Lausanne on November 18-19, 1976. *Communication de l'Institut de Thermique Appliquée de l'EPF de Lausanne*, Nr 5, 1977, pp 133-139.

5.22 Nagao, S. & Sieverding, C.H.: Cascade Tests of Hub and Tip Section of a Last Stage Rotor Blade for a Big Steam Turbine. *von Karman Institute*, unpublished.

CHAPTER 6

Turbine Performance Measurement

6.1 INSTRUMENTATION DEVELOPMENT

M.J. Moore

6.1.1 Introduction

This brief review covers methods of measuring overall turbine performance and developments in instrumentation for the detailed examination of flow fields in the machine.

The drive for higher turbine efficiency necessitates increased accuracy in heat-rate testing to enable small, but highly significant, improvements in performance - and reductions in fuel costs - to be detected. The methods are required by manufacturers for the development of turbines of superior performance in the face of intense international competition for orders. Utilities need reliable acceptance procedures, accurate cost/benefit assessment of retrofits for performance improvement and on-line systems for optimum plant operation.

Detailed information on individual stage efficiencies and the flow distributions within the turbine cylinders is required by research and development engineers to locate areas of high loss and to validate theoretical flow prediction , and design philosophies. Particular areas of interest are in the wet-steam stages where lack of instrumentation to measure steam wetness fraction has, until now, been a serious disadvantage.

Advances in this area are likely to be rapid over the next few years. The advent of cheap and reliable solid-state electronics, the mini-computer and the laser has enabled more measurement to be made in less time by fewer people, such that the costs involved in measuring plant performance are a relatively minor fraction of the costs of effecting improvements.

6.1.2 Measurement of Steam Plant Efficiency

The measurement of heat-rate of turbine condenser and feed-heating system is an established, well documented procedure for which national and international codes have been devised [6.1-6.3]. To place new developments in perspective it is relevant to consider the basic heat-rate measurement process.

Heat rate is defined as the heat input to the cycle per unit electrical power produced. The electrical output, P, of the alternator can be measured conventionally by applying wattmeters, ammeters and voltmeters to each phase and represents probably the least contentious item of the test. Measurements of heat input is more difficult.

The measurement is made by determining the water/steam enthalpy rises in the evaporator-superheater and reheater sections of the boiler as shown diagrammatically in figure 6.1.1. Thus the heat rate, H.R. is :

$$\text{H.R.} = \frac{m_2(h_2-h_1)+m_4(h_4-h_3)}{P}$$

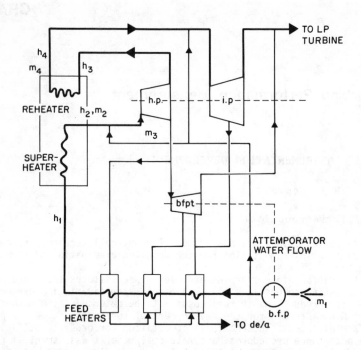

FIGURE 6.1.1   Heat input to a superheat-reheat steam cycle

and whereas enthalpies can be obtained relatively accurately from conventional pressure and temperature measurement, mass flows m_4 and m_2 cannot usually be obtained directly. Installation of orifice plates or venturi-meters in high pressure, high temperature lines and removal for recalibration is expensive, as is calibration at the appropriate Reynolds number. Main condensate flow m_2 is therefore usually obtained indirectly by measurement of flow m_1 earlier in the condensate line where pressures and temperatures are lower[†]. This flow is then corrected for quantities extracted or added, e.g. heater drains, to obtain the flow m_2 entering the boiler.

Reheater flow m_4 is obtained from m_2 after deducting steam flow m_3 to heater and boiler feed pump turbine. Flow m_3 can be measured to acceptable accuracy on a flow meter in the smaller line to these auxiliary plants.

The Codes outline the several small flows which may also require measuring or estimating to obtain the basic as-run heat-rate and describe the chief sources of error such as inadequate isolation of the plant, by-passing through heater tube or valve leaks, etc.

Test conditions invariably deviate from design values and the Codes specify the limits for permissible deviation within which simple correction procedures may be applied with acceptable accuracy. Using these procedures the measured cycle can be converted to the design or reference conditions and the Corrected Heat Rate calculated.

The correction procedure requires the efficiencies of the individual sections of the turbine to be determined from the test heat balance. For this purpose

---

[†]In the UK the measurement is usually made in the downcomer from high level deaerators, whereas in the US, the deaerator inlet is often selected

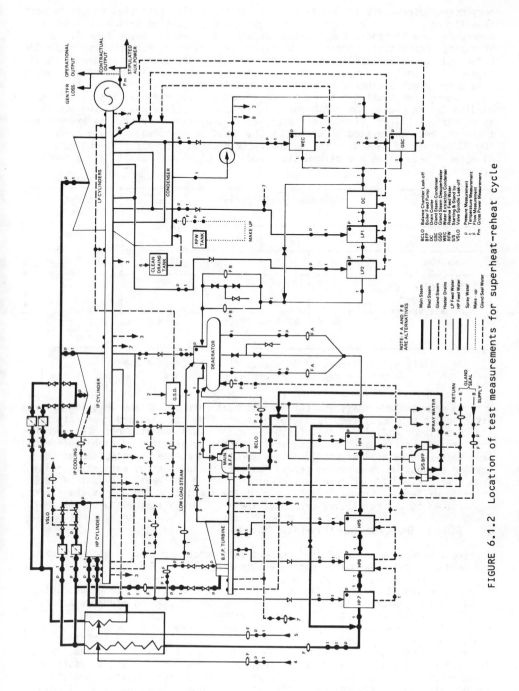

FIGURE 6.1.2 Location of test measurements for superheat-reheat cycle

169

many more measurements are required than for the basic heat-rate, as can be seen from figure 6.1.2 and elaborate data logging and processing systems are now in widespread use to handle the 120-150 measurements, see for example [6.4,6.5]. Similarly mathematical models of the plant are available, e.g. [6.6,6.7] for the rapid and convenient application of the correction procedures. Automated acquisition of data enables many sets of readings to be obtained giving more representative mean values and hence reducing the uncertainty of the final results.

However, when using heat-rate data to detect, for example, the effects of blading design changes on efficiency, it is appropriate to examine the sensitivity of the calculated cylinder efficiencies to measurement error. An example has therefore been calculated by Keeley [6.8], based on an actual, single reheat 500 MW coal fired unit depicted in the cycle diagram figure 6.1.3. Computations were made using the mathematical model of [6.6] and results are shown in Table 6.1.1.

	Cylinder Efficiencies (%)			Cycle Efficiency (%)	Exhaust Wetness (%)
	HP	IP	LP		
As run	84.6	87.1	73.7	39.2	3.5
Effect of error in main feed flow measurement + 1% - 1%	84.6	87.1	70.96 76.13	38.8 39.6	2.35 4.25

TABLE 6.1.1   Effect of main feed flow measurement error
on deduced LP turbine performance

We may conclude that, whereas overall heat-rate and HP and IP cylinder efficiencies may be derived to useful accuracy from the standard heat rate tests, the LP cylinder efficiency calculation is not so well conditioned. Improvements in LP turbine performance by blade design cannot be assessed to high confidence from heat-rate data. The problem stems from the presence of wet-steam at the turbine exhaust which does not permit 'direct' measurement of exhaust enthalpy from pressure and temperature measurement. Instead this enthalpy must be determined by difference from a heat balance of the complete cycle and is therefore susceptible to the cumulative errors of the $\sim$ 100 or so test measurements.

6.1.3   On-Line Performance Monitors

In recent years the value of on-line heat rate measurement for station operators has been realized. Such information enables the utility to :
(i)   schedule plant in a merit order during peaks in demand;
(ii)   generate at maximum efficiency through the operating range of the unit;
(iii)   monitor deterioration in plant performance and plan maintenance accordingly.
To maintain the extensive set of instrumentation required for the full heat-rate test is expensive and simplified procedures are being devised. Figure 6.1.4 shows the more modest instrumentation required by the Simplified ASME test [6.9]. The measured data will give overall as-run heat-rate but do not provide the detail for production of a Corrected Heat Rate. However, for the purposes of plant scheduling (item (i) above), the requirement is for reliable comparisons between the as-run performance of the different plant on the utility system. Hence totally consistent instrumentation, calibrations and installations are necessary, preferably monitored by a single team under the auspices of the

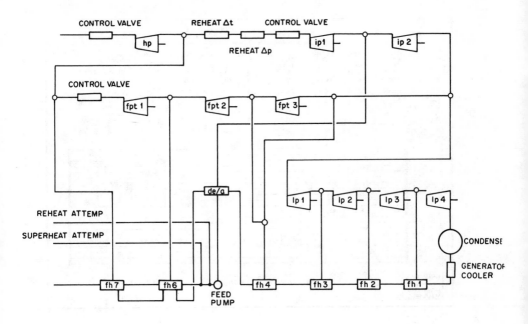

FIGURE 6.1.3   Mopeds network for 500 MW coal fired steam cycle

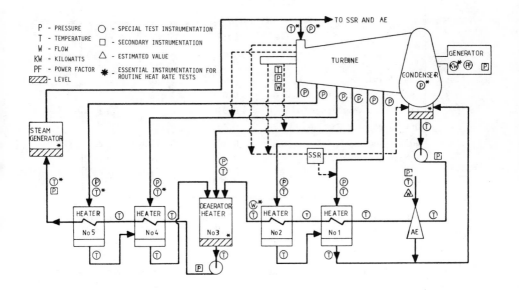

FIGURE 6.1.4   Instrumentation required for ASME simplified performance tests
condensing turbine, regenerative cycle, superheated inlet steam

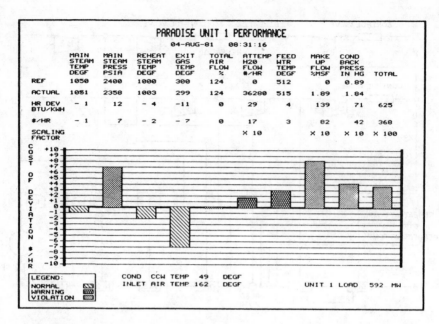

FIGURE 6.1.5a   Control room display at TVA Paradise Power Station

| IRONBRIDGE EFFICIENCY MONITOR UNIT 2 | | | |
| PERFORMANCE DATA | | | |
	EFFICIENCY LOSS %	OUTPUT LOSS MWs	COST PER SHIFT £
T.S.V. Temp Dev	0.027	0.346	43
Reheat Temp Dev	-0.128	-1.681	-207
Turbine Exhaust Pressure Dev	0.091	1.198	147
Final Feed Temp + (Htr Outage)	0.039	0.513	63
Dry & Wet Flue Gas Loss	0.309	4.101	505
CO in Flue Gas	0.001	0.014	2
Reheat Spray Flow Loss	0.111	1.467	181
TOTALS	0.449	5.958	734

Each plant input is scanned 10 times during a 5 minute period, the appropriate calculations are carried out and displayed as shown. The display is held for 5 minutes before changing to the next 5 minute display.

FIGURE 6.1.5b   Control room display at CEGB Ironbridge Power Station

headquarters organization.

For item (ii), the operation of the unit for maximum efficiency, the approach being adopted [6.10] is to process the measured data for presentation in the main control room. The format may typically show the financial penalties or gains to be obtained by control of key parameters, such as reheat temperature or condenser pressure. Figure 6.1.5 shows the VDU in a UK and a US power station. Inability of the operator to meet target settings has obvious implications for (iii), the planning of plant maintenance.

Some simple automatic computation is required to prepare the data for control console presentation. However, the information on individual plant components remains fairly limited and studies are continuing to improve the computer modelling of the plant. The combination of key plant measurements and the mathematical model may then provide detailed information on the performance of individual plant components indicating where deterioration or malfunction is occurring. Such items may include :
- excessive air leaks or tube fouling in the condenser;
- tube leaks or leaks in by-pass valves in the feed heating system;
- seal wear in the turbine.

The plant model may also enable optimization of plant operation during enforced off-design operation during, say, isolation of a feed-heating line due to leaks or part-loading during condenser tube cleaning. Further information on the state of such plant model developments is given in [6.7,6.11,6.12].

6.1.4 Heat Rate by Heat-Rejection Measurement

Increased interest in on-line heat-rate and the development of new flow-measuring systems has encouraged the development of methods of determining heat-rate by measuring electrical power output and the total heat rejected from the cycle. Then, in dimensionless terms :

$$\text{heat rate} = 1 + \frac{\text{total heat rejected}}{\text{electrical power output}}$$

The condenser forms the heat sink of the cycle and measurement of the cooling water heat pick-up provides the total heat rejected. However, the accurate determination of the large CW flow rates and quite small temperature changes is difficult and has discouraged this apparently simple approach in the past.

In [6.13], the measurement of flow rate has been demonstrated successfully on a power station using an ultrasonic 'time-of-flight' technique. A diagram of the installation in a CW outlet pipe is shown in figure 6.1.6, which used 8 transducers, transmitting and receiving on 4 chordal paths. The reference gives the details of the installation which in principle can measure both flow rate and mean temperature. In practice, temperature was measured conventionally using platinum resistance thermometers in a suitable array to provide a representative mean temperature of the stratified outlet water flow.

Results obtained are impressive and support this alternative approach to on-line heat rate measurements. A comparison with conventional heat rate test results is shown in figure 6.1.7a, and figures 6.1.7b and 6.1.7c demonstrate the potential use of the system as a monitor of as-run efficiency.

6.1.5 Direct Measurement of LP Cylinder Efficiency

Various devices for measuring the enthalpy of the partially condensed steam at the turbine exhaust have been under development over several years; a review of the main contenders is given in [6.14]. However, recently the optical methods appear to be the most successful and one of these, described in [6.15 & 6.30], has now been shown to give accurate, repeatable data. However, this system is not yet in widespread use and a brief description will be given below to show the scope of the method.

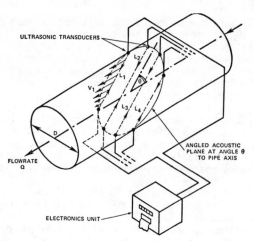

FIGURE 6.1.6 Diagram of ultrasonic flowmeter installation

6.1.5.1 Basis of the Method

Optical measurements in many LP turbines of varying design have shown that the fog droplet mean diameter at turbine exhaust is generally in the low sub-micron range ($0.2 < d < 0.5$ μm). Droplet size can therefore be measured by optical techniques based on Mie scattering. As both droplet size and concentration (wetness fraction) are required the measurement of spectral extinction is appropriate. This technique has found many applications as reviewed by Bayvel & Jones [6.16] but the problem has been to adapt the method for use in steam turbines with their attendant difficulties of access to steam space, whilst maintaining measurement accuracy.

The approach taken by Walters is described in [6.15], the optical system being incorporated into a 25.4 mm diameter probe for use in existing access tubes installed in several operating turbines in the UK for experimental purposes. The general arrangement of the probe is shown in figure 6.1.8, which is fundamentally the same as in [6.15], but has seen several improvements in detail.

The probe is inserted into the turbine and the slot or eye at the inner end is aligned with the steam flow, for example by setting the probe at a yaw angle obtained immediately beforehand by a conventional aerodynamic probe. Monochromatic light, obtained from a white light source via one of a series of interference filters, is transmitted through a flexible quartz-fiber light guide to the probe eye. The light beam is directed across the steam suspension using the optical system shown and detected by the photocell at the extreme end of the probe.

The quantity to be measured by the probe is the transmittance (I/I_0) of the suspension where I, I_0 are the incident fluxes on the photocell with and without the presence of the suspension of water droplets. An essential part of Walter's design is therefore the excluder tube which slides across the probe eye. The light beam passes through this tube to the photocell and steam and water droplets are removed by a flow of purge air enabling the reference intensity, I_0, to be recorded. Retraction of the excluder tube allows the wet-steam flow to be re-established through the slot to give the intensity, I, transmitted through the suspension for the particular wavelength.

Since the publication [6.15], advantage has been taken of new proprietary optical components to widen the spectral range of the system. The second main area of development has concerned the interpretation of results for the polydisperse droplet suspension occurring in LP turbines. The problem of selection

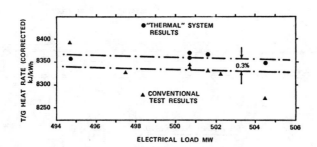

a) Heat rejection of 'thermal' measurement compared
with conventional heat rate test results

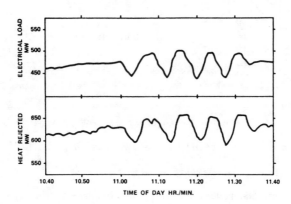

b) Measured variation of heat rejected with load

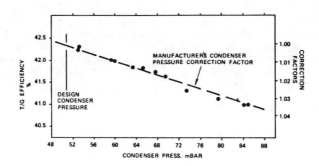

c) Measured effects of condenser pressure on heat rate

FIGURE 6.1.7 Examples of application of heat rejection or 'thermal' measurement

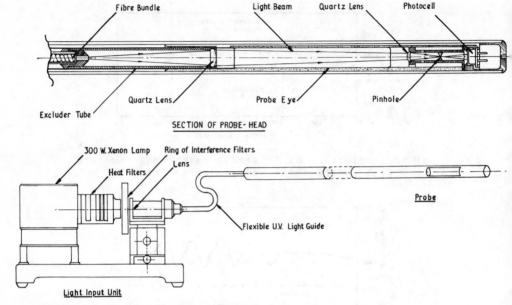

SECTION OF PROBE-HEAD

Light Input Unit

FIGURE 6.1.8 General arrangement of CEGB optical probe

of the best method data processing has been discussed by Walters in [6.17].

6.1.5.2 Interpretation of the Optical Data

For a monodispersion, the reduction in light flux by scattering can be integrated
to give the wall known Bouger law for the transmittance :

$$\frac{I}{I_0} = \exp\left(- nAEs\right)$$

for a suspension containing n droplets per unit volume, each of frontal area A,
for a total path length of s. Extinction coefficient E is a function of the par-
ticle size parameter k $(= \overline{\Lambda}d/\lambda)$ and calculable from the Mie theory for the interaction of an electromagnetic wave on a di-electric sphere. Expressed in terms
of wetness fraction Y, assuming no slip between phases, the Bouger law may be
written :

$$\frac{I}{I_0} = \exp\left[- \frac{3s}{2} \frac{E(\lambda,d)}{d} \left(\frac{Y}{1-Y}\right) \frac{\rho_g}{\rho_f}\right]$$

Experimental studies using a single wavelength laser light source, for
example [6.18], use estimated values of wetness fraction Y to determine approximate mean fog droplet diameter, d. But, if wetness fraction is not known, it is
necessary to obtain a series of transmittance values for a range of light wave-
lengths, λ, to obtain both wetness Y and droplet diameter d.

For a monodispersion a curve matching process can be used by plotting the
experimental results as $\ln (I_0/I)$ versus ν $(= \overline{\Lambda}/\lambda)$ on log-log axes. Taking logs
of the Bouger equation,

$$\log \ln \left(\frac{I_0}{I}\right) = \log E + \log \phi$$

where

$$\phi = \frac{3}{2} \left(\frac{s}{d}\right) \left(\frac{Y}{(1-Y)}\right) \frac{\rho_g}{\rho_\ell}$$

And hence the experimental curve is displaced from the extinction coefficient ordinate by an unknown constant value ϕ. Similarly the abscissa may be expressed

$$\log(\nu) = \log k - \log d$$

As shown diagrammatically on figure 6.1.9, fitting the experimental curve to the theoretical Mie curve of $E(k)$ enables displacements ϕ and d to be obtained (and hence wetness fraction from the above definition of ϕ).

However, for a polydispersion subjected to a monochromatic source of wavelength, D, droplets of different diameter will exhibit different extinction coefficients, E. The resulting transmittance therefore takes the form of an integral equation :

$$g = \frac{1}{t} \ln \left(\frac{I_0}{I}\right) = \int_{d_1}^{d_2} E(d) \, f(d) \, dd$$

where the frontal area of the droplets in the size interval dD is included as :

$$f(d) = \frac{\Lambda}{4} \, C_n \, N_r \, (d) \, d^2$$

where C_n is the total number of droplets per unit volume and N_r (d) is the probability frequency for droplets of diameter d.

Thus an equation for g can be written for each wavelength of light used, the set forming a series of numerical integrations, thus :

$$g_j = \sum_{i=1}^{n} w_i \, E_{ji} \, f_i$$

where w_i are the coefficients of the quadrature formula used (e.g. Simpson's Rule).

In matrix form the above set of equations may be written :

$$\underline{g} = K \cdot \underline{f}$$

which in principle can be inverted directly to give the size spectrum :

$$\underline{f} = K^{-1} \cdot \underline{g}$$

However, the system has been found ill-conditioned, the inevitable small scatter in the experimental data, g, resulting in widely oscillating solutions for f. Walters [6.17] has investigated various alternatives for interpreting the optical data and recommends the method of Phillips and Twomey [6.19,6.20].

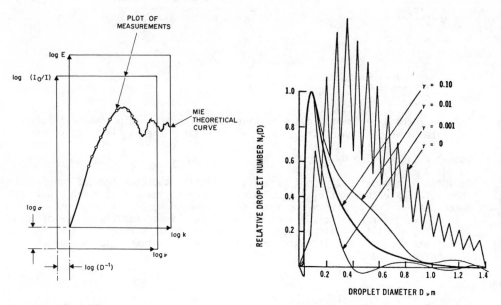

FIGURE 6.1.9 Interpretation of extinction measurements for a monodispersion

FIGURE 6.1.10 Example of effect of smoothing parameter γ on calculated droplet spectrum

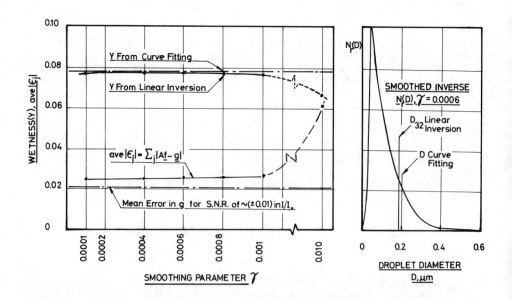

FIGURE 6.1.11 Extinction data inversion: effect of smoothing parameter on deduced wetness fraction

This approach applies the direct inversion but imposes a constraint on the solution by smoothing, that is by minimizing the mean square curvature of $\underline{f}$. An example of a smoothed solution is shown in figure 6.1.10, the smoothing parameter γ being selected to be just sufficient to dampen oscillations without distorting the droplet size spectrum solution. Integrating the spectrum provides wetness fraction and it is found that the result is not sensitive to the value of γ used (Fig. 6.1.11). The mean residual of the numerical inversion is also shown to minimize at a value of the same order as the estimated experimental error in the data, g, suggesting that the inversion process need be refined no further.

6.1.5.3 *Application to Determination of End Point Enthalpy*

As the result of radial variations in efficiency and work extraction in the several stages of the LP turbine, the exit wetness and enthalpy are also not uniformly distributed. To obtain representative mean values an integration of the flow plane is required, for example, mean static enthalpy, $\overline{h}$, is given by

$$\overline{h} = \frac{\oint (h_g - Y h_{fg}) \dfrac{C_z \, \rho_g}{1-Y} \, dA}{\oint \left(\dfrac{C_z \, \rho_g}{1-Y} \right) dA}$$

where C_z is the mixture velocity normal to the plane of integration, A, crossing the flow.

Downstream of the final stage the flow enters the exhaust system and the distribution of velocity and wetness becomes decidely asymmetrical. The simplest measurement is therefore the traverse a few centimeters downstream of the moving blades where axisymmetry may be assumed with probably little error. In practice therefore mean enthalpy is obtained from a single radial traverse at such a position, shown for example in figure 6.1.12.

The resolution of the probe is determined by the length, t, of the slot, the path length of the light beam in the droplet suspension. In the single path system used a path length of order 12 cm is required to obtain an accurately measurable extinction over most of the spectral range for LP turbine exhaust wetness fraction < 0.1. As shown in figure 6.1.12 this resolution is quite adequate to obtain a detailed wetness profile. Conventional aerodynamic probes provide velocity and pressure (hence density ρ_g) distributions enabling the mean enthalpy to be calculated.

A small empirical correction is currently made to the measured wetness distribution to allow for the coarse water invariably present. The optical system will not detect the small quantity of water which has deposited on blading during passage through the final stages and is re-entrained in the flow as larger droplets of > 10 μm diameter. The wetness fraction obtained from the optical probe is therefore multiplied by a correction factor 1.08, based on the catch-pot measurement of coarse water of Williams & Lord [6.21].

A second small correction is made to relate the sample flowing through the probe eye to the undisturbed wet steam mixture - to compensate for the small degree of non-isokinetic sampling and slot end offsets. A Mach number dependent correction factor on wetness fraction has been determined from calibrations in the CERL wet-steam tunnel, the multiplier being < 1.1 for the Mach number range typical of an LP turbine exit plane.

The probe enthalpy measurements are related to the end point of the turbine condition line, computed from a heat rate test, as shown in figure 6.1.13. Turbine exit plane pressure is usually above the reference condenser pressure used in the heat rate computation. However, the mean total enthalpy of the steam remains constant through the exhaust system where no energy is added to or

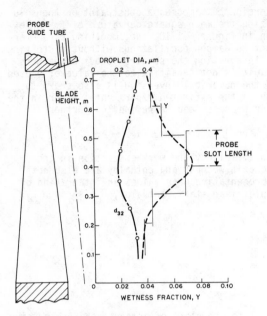

FIGURE 6.1.12 Typical optical probe data from turbine exit plane traverse

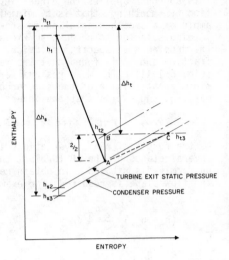

A — STATIC CONDITION AT TURBINE EXIT PLANE
B — STAGNATION CONDITION AT TURBINE EXIT
C — STAGNATION CONDITION AT CONDENSER

FIGURE 6.1.13 LP expansion end point

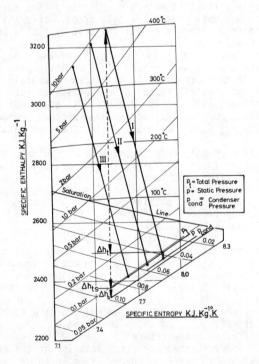

FIGURE 6.1.14 Condition line end point enthalpies measured by optical probe in a 550 MW turbine

extracted from the flow. Hence the quantity for comparison with the heat rate end point enthalpy is the mean total enthalpy at the exit plane from the turbine.

$$\bar{h}_t = \frac{\displaystyle\int_{hub}^{tip} \left[h_g - Yh_{fg} + \frac{C^2}{2} \right] \frac{C_z \, \rho_g}{1-Y} 2\pi rdr}{\displaystyle\int_{hub}^{tip} \left(\frac{C_z \, \rho_g}{1-Y} \right) 2\pi rdr}$$

The optical probe has been used in several turbines in the UK, a turbine in the US and in the Ansaldo machine at Vado Ligure. Fog droplet mean diameter (d_{32}) at turbine exhaust has been found to vary significantly from one turbine design to another in the range $0.10 < d < 0.5$ μm and, of course, wetness fraction and cylinder efficiencies differ also. Agreement with heat rate test data has been within the potential inaccuracies of the latter for LP performance measurement. A demonstration of the consistency and reliability of the optical probe data is shown in figure 6.1.14 where progressive decrease in reheat temperature produced a matching measured increase in exhaust wetness.

6.1.6   Single Stage Performance Measurement

To improve the performance of the wet steam stages, the development engineer requires detailed measurements of both wetness fraction and aerodynamic parameters. Such investigations are often carried out in the laboratory in scale-model turbines using air, Freon/air or superheated steam as the working fluid, the single phase medium allowing enthalpy determination from pressure/temperature data. Model testing allows free access for detailed measurements but has some major disadvantages :
- the effects of scale on the performance of wet steam turbines is still not known;
- reduction in scale reduces the permissible tolerances on blade-section dimensions with attendant high manufacturing costs;
- corresponding miniaturization of instrumentation may not be possible and proportionately large probes may produce spurious data by distorting the flow field;
- model turbines are expensive and long lead times are required for design, construction and commissioning.
Hence measurements in full-size machines are extremely valuable, even though problems of access tend to limit the amount of detail which can be obtained. A typical installation in a CEGB LP turbine, as shown in figure 6.1.15, is limited to single, usually near-radial traverse lines which are positioned, for the final stage, before and after the stage and between fixed and moving blades.

6.1.6.1 Aerodynamic Measurements

Some of the various types of probes in use are reported in [6.14] and below are described the problems currently being addressed. These include the difficulties of measuring transonic flows, the assessment of errors due to flow unsteadiness and the scope for application of laser anemometry.
The final stages of large LP turbines contain substantial regions of transonic flow ($0.85 < M < 1.15$), a notoriously difficult range for instrusive instrumentation. Near-normal shock waves ahead of probe heads and support stems can impinge upon nearby flow boundaries (blade surfaces) and reflect back across the probe or in other ways modify the flow field ahead of the probe sensing point.

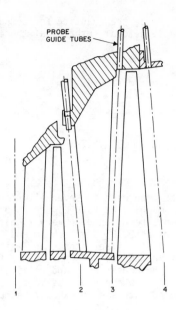

FIGURE 6.1.15   Typical probe traverse positions in
the final stage of a CEGB turbine

Hence probe response may be a function of the local flow field and conventional
probe calibration methods may not be applicable. The task of the researcher is
therefore to develop probes which are less susceptible to such effects by suitable selection of head and stem geometry. The investigation requires a transonic
calibration facility and developments continue in this area.

     For transonic calibration in wet-steam, the well established ventilated wall
techniques used in large transonic wind tunnels have been adapted by Wood [6.22]
for the CERL steam tunnel. The system, shown in figure 6.1.16 has operated
successfully and as a result new slender probe geometries are being developed
for use in the exit region of turbine final stage fixed blades.

     Flow unsteadiness at stator and rotor exit will comprise both periodic com-
ponents at blade passing frequency and a random turbulent background [6.23] .
These fluctuations may corrupt pressure measurements by :
- fluid resonance in the cavities of the pressure measuring system giving rise
to non linear effects;
- non linear aerodynamic effects around the probe sensing head due to fluctuations in main flow velocity and direction.

     Generally the pressure measuring transducer is small and connected to the
sensing head by long, small diameter tubes which can be shown [6.14] to damp out
cavity resonances by frictional resistance. The question of probe head non linear aerodynamics remains, stimulated by an off-quoted conclusion by Samoilovich
& Yablokov [6.29] that Kiel type probes indicate spuriously high pressure in
unsteady flows.

     Experiments using an unsteady flow generator developed for the simulation
of turbine-type unsteadiness have been reported by Wood & Langford [6.24] . Errors
were apparently negligibly small at turbulence levels of < 20%, but results were
considered not sufficiently accurate to draw firm conclusions. Experimentation

is difficult in this area, particularly in respect of obtaining the true, reference pressure and of generating a representative unsteady flow, but further work is undoubtedly required.

Non-intrusive systems using laser anemometry may ultimately be the solution in the difficult transonic areas, but the problems of wet-steam and limited access has so far prevented their application in full size turbines. Nevertheless, the spectacular results obtained in gas turbines, e.g. [6.25,6.26], encourages the application of the technique and systems for LP turbine use are being developed, e.g. [6.27].

6.1.6.2 Stage Efficiency Measurement

During the period of development of methods of measuring wetness fraction, an alternative interim approach was developed for the measurement of stage efficiency from pressure probe data. Known as the momentum method [6.28], this technique may still complement the wetness/enthalpy measurements now being produced by the optical probe for reasons described below.

Summarizing briefly the basis of the momentum method as it is now being applied, the method is based on the Euler turbine equation for the enthalpy drop across a stage in terms of the change in angular momentum flow across the rotor. Thus the classical equation for a simple '2-D' stage, is :

$$\Delta ht = \Omega r \left(C_{\theta 3} - C_{\theta 4} \right)$$

which, as shown in [6.28], can be expressed to good approximation for a wet steam stage as :

$$\Delta h_t = \Omega r \left[C_{\theta 3s} \left(1 - \xi_F \right)^{1/2} - C_{\theta 4} \right] - \Delta h_{ts} \Sigma \zeta$$

where ξ_F is the fixed blade aerodynamic loss coefficient and $\Sigma \zeta$ represents the summation of wetness losses. The application of the method requires the determination of $C_{\theta 3s}$, $C_{\theta 4}$ and Δh_{ts} from pressure measurements. Loss coefficients ξ_F and ζ can be shown to have a small effect on the value of Δh_t so approximate estimates may be used without serious loss of accuracy to the total-to-total efficiency :

$$\eta_{TT} = \frac{\Delta h_t}{\Delta h_{ts}}$$

When applying the method to a typical final stage the above two dimensional equation must be replaced by a through flow computation, linking corresponding points through the stage by stream tubes containing equal increments of mass flow.

However, the practical difficulties of making representative measurements of $C_{\theta 3s}$ at plane 3 are several since the velocity field shortly downstream of the fixed blades is not axisymmetric and, in the transonic and supersonic regions in the lower 1/2-2/3 of the annulus, circumferential variations in pressure and flow angle may be significant. Because only a single traverse line is available, therefore, at plane 3 the measured static pressures and yaw angles are discarded except at subsonic regions and replaced by distributions computed by the through flow method constrained to fit the subsonic region and the measured distributions upstream and downstream of the stage.

The method can therefore be considered a close combination of through flow theory and measurement which maximizes the information obtained from the measurements.

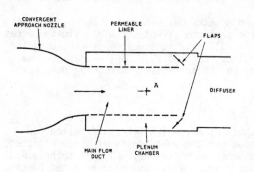

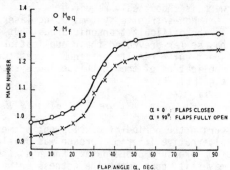

a) Configuration of CERL
   transonic test section

b) Variation of Mach number
   at A with flap angle
   eq. - equilibrium
   f   - frozen flow definition

FIGURE 6.1.16   A ventilated transonic calibration facility for the wet
steam tunnel at CERL

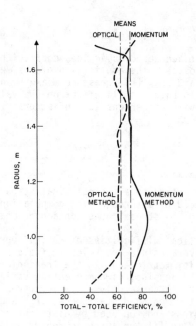

FIGURE 6.1.17   Radial variation in final stage efficiency measured
by two different methods

It is interesting therefore to compare recent results for a single stage
where both momentum method and wetness-enthalpy measurements are available. The
stage in question was from an older machine (long since superseded) but is of

interest to the development engineer in being of poor efficiency. However, as results show (Fig. 6.1.17), whilst arriving at approximately the same figure for overall stage efficiency, the two approaches produce substantially different radial variations in the efficiencies along individual stream tubes.

From the accompanying aerodynamic measurements, the distribution of adverse incidence angles supports the momentum method conclusion of a fall off in efficiency at the tip. Optical probe data in this region, applied in the conventional way to calculation of stream tube efficiency, produces suspiciously high efficiencies in the tip region. Further measurements and analysis is clearly required, but first conclusions are that some radial movement of water relative to the vapour phase may be occurring. Thus the overall efficiency can be obtained by optical probe, but the momentum method may more correctly reflect the radial distribution of losses.

6.2 FIELD MEASUREMENTS IN LP CYLINDER OF A 320 MW TURBINE

A. Accornero & L. Maretto

6.2.1 Introduction

To predict both the overall performance and the exactness of the fluid dynamic design of low pressure steam turbines, it is necessary to have confidence in the results obtained by complex three dimensional computational methods. This confidence can be obtained only if there has been thorough validation of the theoretical results against the experimental data.

In order to get the most complete information about both the steam fluid dynamic behaviour inside the turbine interstage passages and the actual turbine performances at different load conditions, an exhaustive series of tests has therefore been carried out in the low pressure cylinder of an Ansaldo 320 MW turbine, at the ENEL Power Station of Vado Ligure. Several test campaigns led to a complete set of data concerning the operation of this turbine which have then been compared with the performances predicted in several conditions by a computer program of the through-flow type.

After the first, encouraging results [6.31,6.32], a major emphasis has been placed on the analysis of wetness development, with the use of special probes for direct wetness measurements in the last stages of the low pressure turbine, and the allowance for condensation shock effects in the computing program. In this paper only the most relevant points concerning these aspects of the work will be dealt with, while most detailed information about both the experimental and theoretical aspects of the whole work can be found in [6.31,6.33,6.34].

6.2.2 The Test Turbine

The Ansaldo 320 MW power plant is built around a tandem-compound, double-casing reheat turbine with an HP-IP cylinder and a double flow LP cylinder. The eight stage regeneration cycle (Fig. 6.2.1) is provided with three low pressure exchangers, a deaerator and four high pressure feed heaters.

The main design characteristics of this turbine are :

power	320 MW	LP inlet temperature	365°C
rotating speed	3000 RPM	LP inlet pressure	9.5 bar
HP inlet flow rate	284 kg/s	condenser pressure	50 mbar
HP inlet temperature	538°C	number of LP stages	6
HP inlet pressure	167 bar	LS blade height	850 mm
RH temperature	538°C		

The casing flare of the low-pressure cylinder flows, resulting in high blade speeds for the last stage and a pressure ratio of over 6:1, coupled with a hub: tip radius of about 2, results in high relative Mach numbers (about 1.7) at last

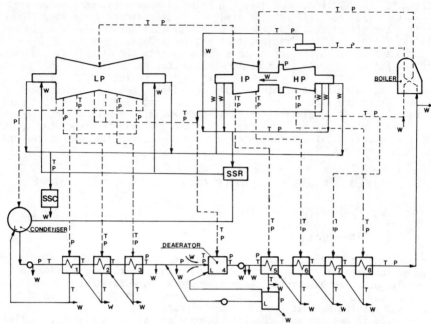

FIGURE 6.2.1 Schematic of 320 MW turbine cycle

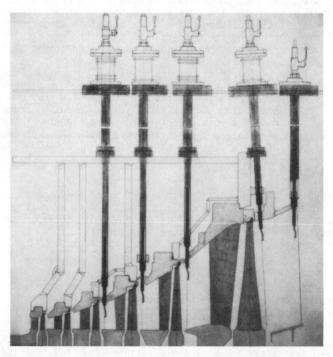

FIGURE 6.2.2 Cross-section of instrumented 320 MW turbine LP cylinder

stage nozzle root and rotor tip.

6.2.3   Turbine and Cycle Setup

The arrangement of the turbine for the purposes of these tests had to fulfil several requirements, imposed mainly by the necessity to not interfere with the normal operation of the plant. During the tests, the plant was used for continuous power generation with allowance only for load variations in strictly limited periods of the day or, in extreme cases, of the week.

Figure 6.2.3 shows how the probe insertion system has therefore been conceived: a tube has been installed through the inner casing with a terminal conical guide for the probe stem near the flow passage, while at the opposite end, it terminates at an expansion joint on the outer casing. A continuous leak tightness during probe insertion and traversing is guaranteed by the combined system of a ball valve at the top of the guide tube and a seal system on the probe carriage. This latter, though mounted on the outer casing, follows the guide tube and inner casing displacements through a bi-directional gliding bearings system.

In this way probe insertion in any of the measuring planes is permitted when the turbine is running whilst, at the same time, avoiding any steam leakage or air intake through the casing. From the mechanical point of view, the differential thermal expansion of inner and outer casings is accommodated whilst providing an accurate guide for the probes.

The measuring planes, shown in the cross-section of figure 6.2.2, are positioned at five axial locations, i.e., at the exit of the third, fourth, fifth, sixth stage and exhaust diffuser. Only at the exit of the sixth stage is the probe inclined to the axis so that the traverse is parallel to the blade trailing edge. At the inlet of first, second and third stages, where no radial measurements were made, temperature and wall pressure measuring points were provided. Furthermore, in all measuring planes where temporarily no moving probes were in use, fixed probes were placed in the middle of the steam passage for the measurement of reference static and total pressures.

The turbine plant was also completely instrumented, according to ASME Performance Test Code [6.2], for the evaluation of heat balance, flow rate distribution in the LP cylinder and turbine stability; in figure 6.2.1 the most relevant measurement points on the cycle are indicated.

6.2.4   Instrumentation

The probes used for aerodynamic measurements were, in the first series of tests, [6.3] of the five-hole spherical head type with a shielded thermocouple incorporated into the stem. This kind of probes, however, due to the relatively large head dimensions, did not give completely satisfactory results when working in high velocity gradients such as at the exit of last stage.

In the last series of tests two different probes were then used; a disc probe (Fig. 6.2.4) with directional holes on the rim for the measurement of yaw and pitch angle and static pressure, and a separate probe with shielded thermocouple and kiel head (Fig. 6.2.5) for the measurement of temperature and total pressure. The advantages of using these probes reside in the high sensitivity of the disc probe to yaw angle variations, so that a precise nulling of the signal for the evaluation of yaw angle could be done, and in its very low error in the static pressure measurement. Furthermore the high intensitivity to flow direction of shielded probes for total pressure and temperature (the probe used in these tests had an insensibility cone of ± 25°) allowed the measurement of these quantities to be done in sequence with that of angles and static pressure, with negligible errors due to flow misalignment.

The aerodynamic probes were calibrated in an open-jet air wind tunnel, by nulling the yaw angle signal, over a Mach number range of 0.4 ÷ 0.9 and a pitch

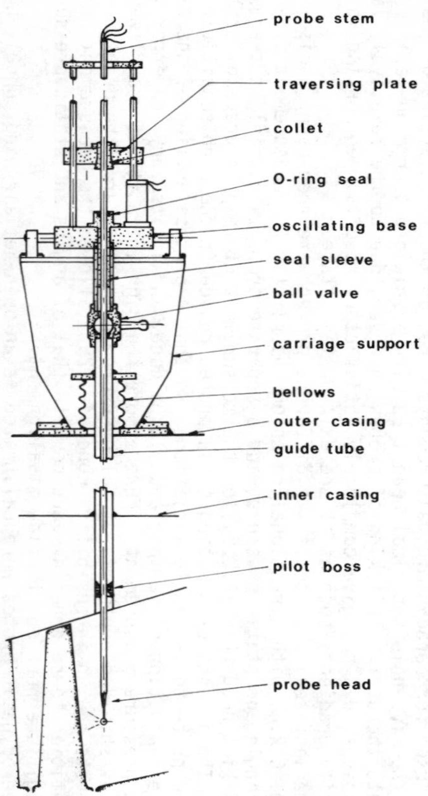

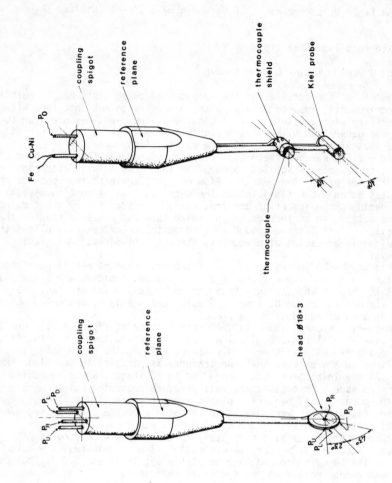

Fe Cu-Ni P_O

coupling spigot

reference plane

thermocouple shield

Kiel probe

thermocouple

FIGURE 6.2.5 Kiel probe for total pressure and temperature

coupling spigot

reference plane

head Ø18×3

P_R

P_D

P_U

P_L

23°

40°

FIGURE 6.2.4 Disc probe for yaw and pitch angles and static pressure

probe stem

traversing plate

collet

O-ring seal

oscillating base

seal sleeve

ball valve

carriage support

bellows

outer casing

guide tube

inner casing

pilot boss

probe head

FIGURE 6.2.3 Probe insertion system

angle range of ±30°, and a subsequent check has been done in a steam tunnel. In order to check the recovery factor, the shielded thermocouple head has also been calibrated in a steam tunnel, both in wet and dry steam.

The direct wetness measurement in the last stages of the LP cylinder has been performed by using an optical probe based on the extinction method [6.17, 6.35] and shown in figure 6.1.8 of section 6.1. The probe was for these tests specially modified such as to cope with the dimensional requirements of the probe insertion system, namely traversing length and outer diameter.

All the pneumatic signals coming from the probe, as well as those coming from the pressure taps on the LP cylinder, were fed to electric pressure transducers, periodically calibrated against dead weight testers. Before each measurement all the pressure lines from the probe heads to the transducers were automatically purged through a solenoid valves system, shown in figure 6.2.6. This

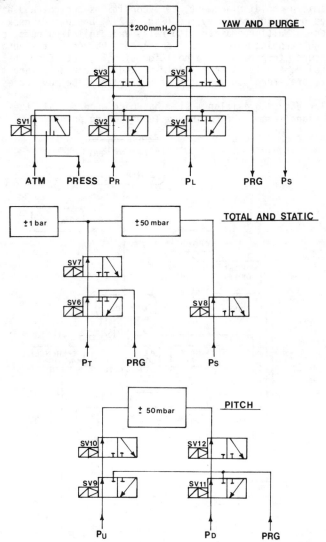

FIGURE 6.2.6 Electro-pneumatic system for transducers purging and nulling

system allowed for periodic transducer calibration, pressure balancing for
transducer offset reading, simultaneous purging of the pressure lines either
with compressed air or, in the last stages, by sucking atmospheric air, and
finally connecting each transducer with its pressure line.

The most critical pressure measurements on the cycle, namely those relevant
to the evaluation of turbine expansion line, have been done by means of high
precision conventional instruments, such as U-tube manometers, dead weight
testers and calibrated Bourdon manometers. All other cycle pressure measurements have been done by means of strain gauge transducers, while thermocouples
with ice point reference junctions have been used for temperature measurements.

6.2.5   Data Acquisition and Control System

The aerodynamic probe displacement, test conditions control and data acquisition
are performed by an automatic system (Fig. 6.2.7) based on a desk-top computer
which, through a Multiprogrammer with relay and digital outputs, controls the
probe carriage stepping motors and the solenoid valves of the purging circuit.
All the pressure, temperature and position signals, both from the probes and
from the turbine, are fed to a multiplexer and A/D converter and then, through
the computer, are stored on magnetic or perforated tape for successive data
reduction.

A similar data acquisition system has been used for all the pressure and
temperature signals coming from the cycle measuring points.

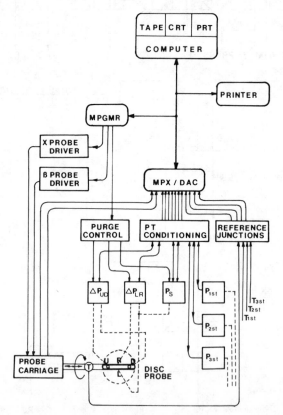

FIGURE 6.2.7 Data acquisition and control system

Due to the complexity of its operation the optical probe was on the contrary manually traversed and operated during wetness measurements.

6.2.6 Measuring Technique

Traverses were made with aerodynamic probes in all of the five measuring planes in several turbine working conditions, namely at full load (330 MW), at 70% and 80% of full load, and at different condenser pressure levels, ranging from 40 to 120 mbar. Wetness measurements were made, at the exit of fourth and fifth stages and of the diffuser, immediately after aerodynamic measurements taken at exactly the same radial positions, at 60%, 70%, 80% and 100% load.

The measuring procedure began after turbine setting at the required working condition, and allowing for due stabilization of thermodynamic conditions. Traverses then started following a measuring cycle including probe displacement, purging of the pressure lines, alignment of the probe with the flow in the circumferential plane, acquisition of signals coming from probes and LP cylinder; this cycle was repeated at least five times before displacing the probe to the next measuring point.

This procedure, due also to the length of the periodical purging and following pressure line stabilization, required on the average from 2 to 10 minutes at each radial position, depending on the flow characteristic, and a complete traverse in the last measuring planes from 6 to 8 hours. The use of an average of five or more sets of readings, however, assured that random errors were negligible.

Wetness measurements, with contemporary use of aerodynamic and optical probes, required at each position a much longer time, around 30 minutes or more, due also to the intrinsic requirements of the absorption method. This characteristic, together with the little time available in the period these tests were carried out, led to a limitation in the number of measured points, consistent, however, with the need of getting sufficiently detailed information about wetness development in the last stages.

Considering then the overall time required for the acquisition of a complete set of data at each measuring plane, the turbine stability was continuously monitored and every five minutes cycle data were acquired as reference conditions for probe data. This was also necessary in order to be able to restart the tests in case of accidental interruption of the measuring points and to make the measurements in all the planes at the same working conditions.

6.2.7 Experimental Results and Comparison with Calculations

Data from aerodynamic probes traverses have been processed to give flow characteristics in the measuring planes; reference steam thermodynamic conditions, particularly for the last stage which is always working in the wet region, came from expansion line evaluations by heat balance calculations, from which also interstage flow rates were derived.

Sample calculations have also been made in all the working conditions for which experimental test results were available, but a major interest has been placed in the influence that the allowance for condensation shock effects had on the estimation of flow characteristics across the last stage. In this paper only the results relevant to the last stage at full power will be considered, as these data are of the most interest in this respect.

The thermodynamic conditions chosen as input data for this comparative calculation were those coming from a heat balance evaluation based on cycle measurements taken during the traversing of aerodynamic probes at the exit of the last stage at full power, namely :
- inlet steam conditions to the last four stages
     mass flow rate   101.914 kg/s
     static pressure   3.055 bar
     static enthalpy   2912.0   kJ/kg

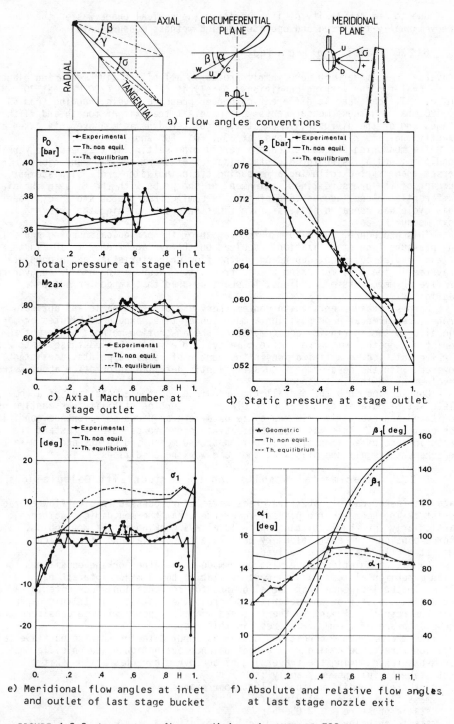

a) Flow angles conventions

b) Total pressure at stage inlet

c) Axial Mach number at stage outlet

d) Static pressure at stage outlet

e) Meridional flow angles at inlet and outlet of last stage bucket

f) Absolute and relative flow angles at last stage nozzle exit

FIGURE 6.2.8 Last stage flow conditions in ANSALDO 320 MW steam turbine

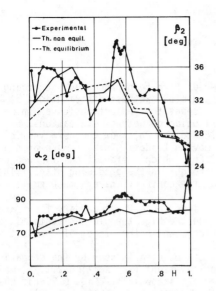

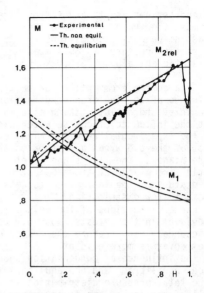

g) Relative Mach numbers at last stage nozzle and bucket exit

h) Absolute and relative flow angles at last stage bucket exit

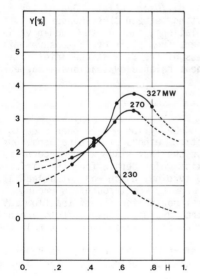

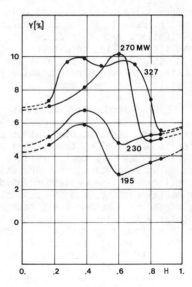

j) Wetness fraction at last stage inlet

k) Wetness fraction at diffuser outlet

FIGURE 6.2.8   Last stage flow conditions in ANSALDO 320 MW steam turbine (continued)

- bleed between fourth and fifth stage
    mass flow rate   8.461 kg/s
- last stage outlet conditions
    static pressure   65.4 mbar
    static enthalpy   2350.0 kJ/kg

Data coming from optical probe traverses have been processed, according to
[6.17], so as to give droplet dimensions and steam wetness distributions at last
stage inlet and outlet, taking also into account the indications given in [6.36]
about the influence of coarse water distribution on wetness evaluations.
     The experimental results (Figs. 6.2.8b to 6.2.8h) show the aerodynamic
effect of the tie wires and stubs on the moving blades. This is more pronounced
in the plane at the last stage inlet (6.2.8b), which suffers the influence of
the tie wire placed at about 60% of blade height in the preceding stage. The
welded stubs of the last stage buckets, on the contrary, have a much smaller
effect of flow distribution.
     Large distortions of the velocity profiles caused by secondary flows are
not evident in these tests, possibly because the high turbulence levels mix out
the secondary flows before the traverse plane, while the effect of the last stage
bucket cover is more evident.
     The influence of condensation shocks can be clearly seen by the comparison
between equilibrium and non equilibrium calculations (see section 2.1.9), mainly
in the total pressure distribution at the last stage inlet (Fig. 6.2.8b), where
equilibrium calculations do not agree with experimental results. In all other
cases the difference, though significant, is not of sufficient magnitude to be
detected from the agreement between theoretical and experimental results.
     In figures 6.2.8j and 6.2.8k, the wetness fraction distributions are shown,
at different loads, at last stage inlet and diffuser outlet. One of the most
interesting results is the difference existing between these measurements and
the evaluations made by heat balance calculations, such as the sudden variation
in last stage working conditions, not evident from plain cycle measurements,
from 230 to 270 MW, and the absolute dryness of the flow at the inlet of the
fifth stage (not shown here), while from heat balance evaluations it appeared
to be still in the wet region.

6.2.8   Concluding Remarks

The results of tests carried out on a 320 MW LP cylinder have been used to check
the confidence level of theoretical prediction methods of turbine behaviour,
mainly as far as flow configurations across the last stage were concerned.
     Some uncertainties, however, still exist which derive from the differences
between the real flow behaviour inside the turbine, which was not a laboratory
experimental machine, and the assumptions made by the throughflow program. The
aging of the machine and the circumferential non uniformity of the flow may be
particularly important at the exhaust of the turbine. These effects are not
taken into account by the calculation program and play an important role when
attempting to predict turbine efficiency.
     More detailed investigations are therefore planned on other machines both
to examine thoroughly these latter aspects of the problem, and to get a larger
amont of useful operational feedback.

ACKNOWLEDGEMENTS

The work herein described has been made possible by the active collaboration of Dr. M.J. Moore
and Mr P.T. Walters from CEGB-CERL, Leatherhead, who personally carried out wetness measurements and evaluation, and gave useful suggestions about aerodynamic probes improvements; of
Mr L. Zanangeli and his staff, from ENEL-DCO/LCP, Piacenza, who took care of all cycle measurements and heat balance calculations, and of the whole personnel of the ENEL power station of
Vado Ligure.

NOMENCLATURE

A   area

C   velocity

d   droplet diameter

E   extinction coefficient

h   specific enthalpy

h_{fg} specific enthalpy of evaporation

H   non dimensional blade height

M   Mach number

m   mass flow

n   number of droplets

P   power

p   pressure

r   radius

s   distance

Y   wetness fraction

α absolute flow angle in circumferential plane

β relative flow angle in circumferential plane

η efficiency

λ wavelength

ξ losses

ρ density

σ meridional flow angle

Ω rotational speed

Suffixes

1   conditions at last stage bucket inlet (Fig. 6.2.8)

2   conditions at last stage bucket exit (Fig. 6.2.8)

D   down

g   vapour phase

L   left

R   right

rel   relative to moving blade

S   static

t   total

U   up

z   axial

θ circumferential

Abbreviations

HP   high pressure

IP   intermediate pressure

LP   low pressure

LS   last stage

RH   re-heat

REFERENCES

6.1   CEGB Steam Turbine Generator Heat Rate Tests, Site Test Code No. 2, 1978.
6.2   ANSI/ASME Steam Turbine Performance Test Code, PTC6-1976.
6.3   VDI Thermal Acceptance Tests of Steam Turbines, *DIN* 1943, 1975.
6.4   Development of a Plant Performance Monitoring System. Paper presented at CEGB Conference on *'On-Line Determination of Thermal Efficiency'*, London, March 1982.
6.5   Golden, G.B.; Pennell, D.J.; Comcaux, J.A.: Major Equipment Testing. EPRI Workshop on *'Fossil Plant Heat Rate Improvement'*, 1981.
6.6   Thompson, M.A.: MOPEDS - Modular Performance, Evaluation and Design System. *CEGB Private Communication.*
6.7   KeHenacker, W.C. & Klink, P.H.: Performance Evaluation Studies. EPRI Workshop on *'Fossil Plant Heat Rate Improvement'*, 1981.
6.8   Keeley, K.R.: *CEGB Private Communication.*
6.9   ANSI/ASME Simplified Procedures for Routing Performance Tests of Steam Turbines, *PTC 65-1979.*
6.10 Holmes, J.G.: TVA's Strategy for Heat Rate Improvement. EPRI Workshop on *'Fossil Plant Heat Rate Improvement'*, 1981.
6.11 Ready, A.B.: The POSTMAN Model for Conventional Power Stations, CEGB Midlands Region Scientific Services, *Private Communications.*
6.12 Legg, J.K. & Price, M.R.: Plant Performance Monitoring Data Program for Georgia Power Company Steam Units. EPRI Workshop on *'Fossil Plant Heat Rate Improvement'*, 1981.
6.13 Sochaczewski, Z.W.; Clay, C.A.E.; Morris, J.A.: Development of a Turbine Generator Thermal Performance Monitoring System. *Proc. Inst. Mech. Engrs.*, Vol. 195, No. 31, 1981.
6.14. Moore, M.J.: Instrumentation for Wet Steam : A Review of Instrumentation for Wet Steam. In: *'Two-Phase Steam Flow in Turbines and Separators'*, Hemisphere, 1976, pp 191-249.
6.15 Walters, P.T.: A Simple Optical Instrument for Measuring the Wetness Fraction and Droplet Size of Wet-Steam Flows in LP Turbines. In: *'Steam Turbines for the 1980's*, The Institution of Mechanical Engineers, London, October 1979, pp 337-348.
6.16 Bayvel, L.P. & Jones, A.R.: Electromagnetic Scattering and its Applications. London, *Applied Science Publishers*, 1981.
6.17 Walters, P.T.: Practical Applications of Inverting Spectral Turbidity Date to Provide Aerosol Size Distributions. *Applied Optics, Vol. 19, No. 4*, July 1980.
6.18 Kreitmeier, F. & Schlachter, W.: An Investigation of Flow in Low-Pressure Wet-Steam Model Turbine and its Use for Determining Wetness loss. In: *'Steam Turbines for the 1980's*, The Institution of Mechanical Engineers, London, October 1979, pp 385-395.
6.19 Phillips, D.L.: *J. Assoc. Computing Machines*, Vol. 9, 1962, p 84.
6.20 Twomey, S.: *J. Assoc. Computing Machines*, Vol. 10, 1963, p 97.
6.21 Williams, G. & Lord., M.: *Third Conference on Steam Turbines*, Gdansk, September 1974.
6.22 Wood, N.B.: A Simple Ventilated Transonic Nozzle for Probe Calibration in Wet Steam. Paper presented at the Symposium on *'Measuring Techniques for Transonic and Supersonic Flows in Cascades and Turbomachines'*, Ecole Centrale de Lyon, October 1981.
6.23 Wood, N.B.: *Inst. Mech. Engrs. Conference Publication* 3, 1973.
6.24 Wood, N.B. & Langford, R.W.: Preliminary Investigations into the Effects of Flow Unsteadiness on the Time Mean Response of a Kiel-Type Total Pressure Probe. Symposium on *'Measuring Techniques for Transonic and Supersonic Flows in Cascades and Turbomachines'*, Ecole Centrale de Lyon, October 1981.
6.25 Gill, M.E.; Forster, C.P.; Elder, R.L.: Measurements in Turbomachines Using Two-Spot Anemometry. In: *Photon Correlation Techniques*, Springer Verlag, 1983, pp 197-204.
6.26 Schodl, R.: A Laser-Two-Focus Velocimeter for Automatic Flow Vector Measurements in the Rotating Components of Turbomachines. *ASME Transact., Series I : Journal Fluids Engineering*, Vol. 102, No. 4, December 1980, pp 412-419.
6.27 Troilo, M. & Maretto, L.: A Dual Focus Fibre Optic Anemometer for Measurements in a Wet Steam Turbine. *'Measuring Techniques for Transonic and Supersonic Flows in Cascades and Turbomachines'*, Ecole Centrale de Lyon, October 1981.
6.28 Moore, M.J.; Wood, N.B.; Jackson, R.; Langford, R.W.; Keeley, K.R.; Walters, P.T.: A Method of Measuring Stage Efficiency in Operating Wet-Steam Turbines. In: *'Steam Turbines for the 1980's*, The Institution of Mechanical Engineers, London, 1979, pp 267-280.

6.29   Samoilovich, G.S. & Yablokov, L.D.:   *Thermal Engineering,* Vol. 17, 1970.
6.30   Walters, P.T.:   Wetness and Efficiency Measurements in L.P. Turbines with an Optical
       Probe as an Aid to Improving Performance.   *ASME Joint Power Generation Conference,*
       Milwaukee, October 1985.
6.31   Accornero, A.; Doria, G.; Maretto, L.; Zunino, E.:   Flow in a 320 MW Low-Pressure
       Section : Theoretical and Experimental Evaluation.   In: *'Steam Turbines for Large
       Power Output',* VKI LS 1980-06, April 1980.
6.32   Hirsch, Ch. & Denton, J.D. (Eds.):   Through Flow Calculations in Axial Turbomachinery.
       AGARD AR 175, October 1981.   ˙
6.33   Accornero, A.; Liberti, V.; Vallarino, G.; Zunino, E.; Zanangeli, L.:   Applicazione del
       resultati della ricerca del Progetto Finalizzato Energetica per i rivievi sperimentali
       sulla turbina de 320 MW di Vado Ligure.   *Giornate di Studio CNR* - Genova, June 1982.
6.34   Doria, G. & Troilo, M.:   Throughflow Calculation on Large Steam Turbine.
       6th Conference on Large Steam Turbines, Pisen, May 1979.
6.35   Walters, P.T. & Skingley, P.C.:   An Optical Instrument for Measuring the Wetness Fraction
       and Droplet Size of Wet Flows in LP Turbines.   *Proc. Inst. Mechanical Engineers,*
       Vol. 141, 1979.
6.36   Williams, G.J. & Lord, M.J.:   Measurement of Coarse Water Distribution in the LP
       Cylinder of Operating Steam Turbines.   *Proc. Inst. Mechanical Engineers,* Vol. 190, No. 4,
       1976.

CHAPTER 7

Low Pressure Turbine Exhaust System Design

H. Keller

7.1 INTRODUCTION

Without any doubt the LP turbine exhaust system is one of the most promising items for improvement in efficiency. The exhaust energy represents more than two percent of the total available isentropic energy and approximately 15 percent of the total losses on large turbogenerators. The exhaust ducts are usually extremely short, curved passages deflecting the flow from the axial to a radial direction and imposing diffusion. They do not, therefore, compare well with conventional straight diffusers. Nevertheless, their performance can be closely related to conventional diffuser theory and, for this reason, a survey of the literature on general diffuser technology is given below.

7.2 A REVIEW OF CONVENTIONAL DIFFUSER DATA

The performance of a diffuser is usually defined by the diffuser efficiency and the recovery factor (Fig. 7.2.1). The diffuser efficiency refers to the gain in static enthalpy as compared to the difference of kinetic energies between diffuser inlet and outlet calculated on the basis of the conditions of continuity. The recovery factor, C_p, is the ratio of gain in static enthalpy to the kinetic energy at the diffuser inlet.

Optimal geometries for rectilinear diffusers, for maximum efficiency, are given by Sovran & Klomp [7.1] and many subsequent publications have been based on their work enabling different diffuser problems to be analyzed on a comparable basis. The main geometrical data (Fig. 7.2.1), the area ratio and the relative diffuser length define performance levels which are quite similar for rectangular, conical and annular diffusers as long as there is no substantial bending of the diffuser (defined as the bend radius/length ratio).

The definition of recovery factor also allows us to compare the area ratios for different Mach numbers which correspond to the same value of C_p (Fig. 7.2.2). The differences in area ratio become considerable as Mach numbers increase beyond 0.4.

Figure 7.2.3 depicts the flow regime chart for two dimensional diffusers [7.1, 7.2] which may be used without significant error for conical and annular straight diffusers as well, if the initial height, h_1, is replaced by the inlet radius (or the difference in inlet radii for annular diffusers). Four different flow regimes exit, as Sovran & Klomp explain. In the region of no appreciable stall flow is steady and uniform, whilst in the region of transitory stall flow is unsteady and non-uniform; flows in the fully-developed stall and jet-flow regimes are reasonably steady, but not very uniform.

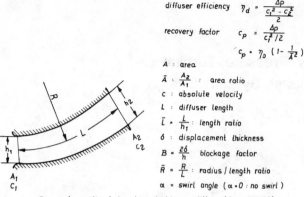

diffuser efficiency $\quad \eta_d = \dfrac{\Delta p}{\frac{c_1^2 - c_2^2}{2}}$

recovery factor $\quad c_p = \dfrac{\Delta p}{c_1^2/2}$

$\qquad\qquad\qquad c_p = \eta_D \left(1 - \dfrac{1}{\bar{A}^2}\right)$

A : area

$\bar{A} = \dfrac{A_2}{A_1}$: area ratio

c : absolute velocity

L : diffuser length

$\bar{L} = \dfrac{L}{h_1}$: length ratio

δ : displacement thickness

$B = \dfrac{2\delta}{h}$: blockage factor

$\bar{R} = \dfrac{R}{L}$: radius / length ratio

α = swirl angle ($\alpha = 0$: no swirl)

Dependence on Re : not significant for steam turbine conditions (Re > 500 000)
Dependence on M : not significant for M < 0.7, for M : 0.7 ... 1.0 : $\bar{A}$-adjustments required.

Dependence on B :

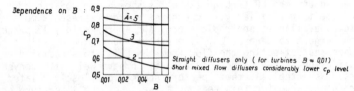

Straight diffusers only (for turbines $B \approx 0.01$)
Short mixed flow diffusers considerably lower c_p level

FIGURE 7.2.1   Some definitions and relations on diffusers

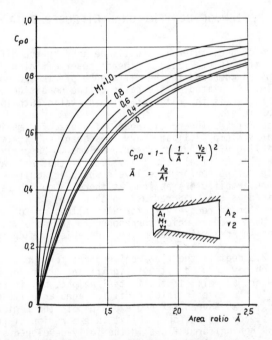

$$c_{po} = 1 - \left(\dfrac{1}{\bar{A}} \cdot \dfrac{v_2}{v_1}\right)^2$$

$$\bar{A} = \dfrac{A_2}{A_1}$$

FIGURE 7.2.2   Theoretical diffusers recovery factor C_{p0} versus initial
Mach number M_1 and area ratio $\bar{A}$ ($\gamma = 1.3$), from [7.1]

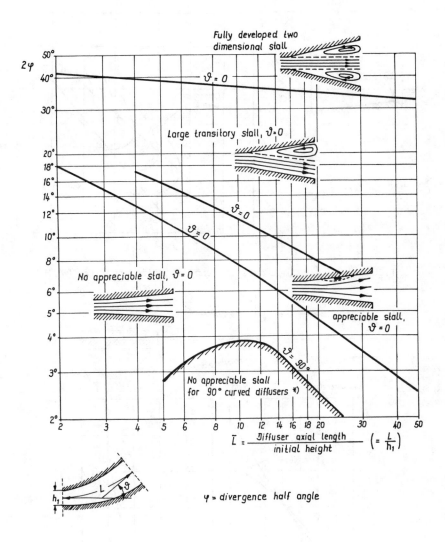

FIGURE 7.2.3 Flow regime chart for two dimensional diffusers, from [7.1]

Further important factors determining diffuser behaviour are the inlet flow swirl angle and the blockage factor, as a measure of the flow velocity profile at inlet. The effect of blockage on recovery factor is shown in figure 7.2.1 and may be used with the performance chart of figure 7.2.4 for conical diffusers [7.1]. The chart includes the loci of optimum performance for given length and area ratios. On the basis of this same reference, Traupel [7.2] has given a computational guide for conversion of this optimization chart for two-dimensional and annular diffusers (Fig. 7.2.5).

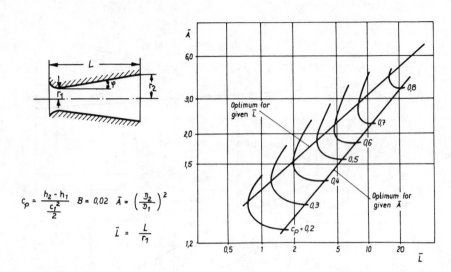

$$c_p = \frac{h_2 - h_1}{\frac{c_1^2}{2}} \quad B = 0,02 \quad \bar{A} = \left(\frac{\vartheta_2}{\vartheta_1}\right)^2$$

$$\bar{l} = \frac{L}{r_1}$$

FIGURE 7.2.4   Optimal data for conical diffusers, from [7.1]

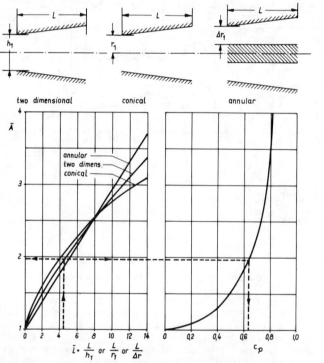

The optimal geometry is fairly independent on blockage factor B

FIGURE 7.2.5   Optimal recovery factors C_p for
diffuser geometries (B = 0.01), from [7.2]

7.3 DATA FOR 90° CURVED DIFFUSER

The performance chart of figure 7.2.3 has been complemented by data for 90° curved diffusers [7.2] which clearly shows that customary exhaust ducts begin to stall at divergence half-angles smaller than those for a non-stalling straight diffuser. A similar performance chart for annular diffusers for gas turbine applications [7.3] is shown in figure 7.3.1. This shows data from experimental studies on the performance of axial-radial diffusers with 90° turn for a defined diameter ratio and comes closer to customary steam turbine applications than the previous chart. But the number of tests seems to have been small and, just in the region of the most interesting radius ratio (below values of 4) and area ratios, the information seems to be contradictory to that given in other publications (which are not charted as they are isolated contributions). Furthermore an optimal 90° mixed flow diffuser certainly does not feature a simple quarter circle contour as shown in this chart.

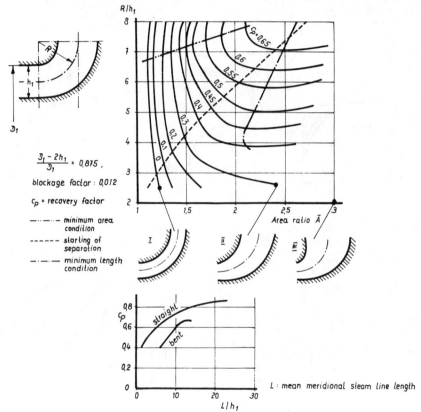

Performance comparison between straight and bent diffuser:
(minimum length condition)

FIGURE 7.3.1   An experimental study of annular diffusers, from [7.3]

Figure 7.3.2 [7.4] shows the results of tests on bent annular diffusers with different contours on the inner and outer walls. The manufacture of large welded LP exhaust hoods requires, to a certain extent, the use of wall elements of conical shape which are welded together to form the pressure surface contour. As may be expected, the influence of this contour on recovery is rather small whereas the

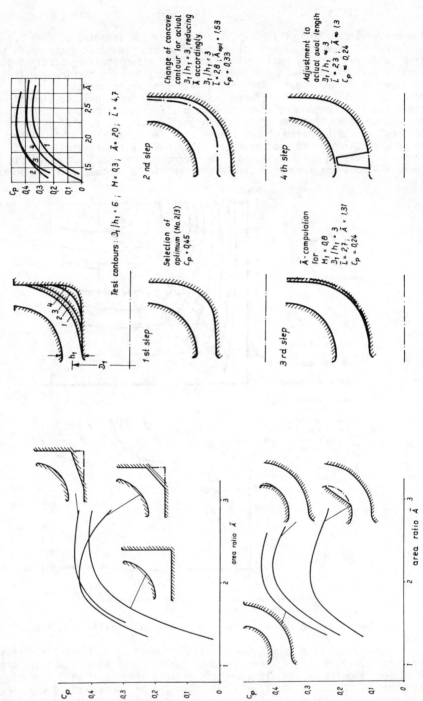

FIGURE 7.3.2 Effect of the shape of axiradial diffusers on the recovery factor, from [7.4]

FIGURE 7.3.3 Comparison of radial annular diffuser tests – actual design data (rough approach)

shape of the suction side is important, particularly in the region close to the diffuser entrance where flow separation must be avoided. Information from reference [7.4] is used in a rather approximate design method, starting from one of the rare quantitative contributions in the literature on the recovery factor for sharply curved mixed flow diffusers with defined contours. The example in figure 7.3.3 shows a step-by-step approach, starting with the selection of the optimum shape with subsequent conversion for diameter ratio, area ratio correction for Mach number and final adjustment to the actual position of the last stage blade.

No publication has been found which compares an annular mixed flow diffuser with a cylindrical flow against a rectangular wall. Diffusers of this kind are often used in air conditioning and their similarity to the customary short mixed-low diffusers in steam turbines is obvious. This offers another possibility of applying existing test data. In figure 7.3.4 a configuration of this kind of non-annular mixed flow is shown, the axial outlet width of which was optimized by Ruchti [7.5] in tests for a given diameter ratio and outer meridional bend. Inserted in this diagram is the inviscid free cylindrical jet flow impinging on a plane. The outer streamline is by definition an isobar of constant velocity. A flow of this kind is not a diffusing flow but is certainly a flow with low loss-level when enclosed in a casing. Assuming streamlines as passage boundaries and applying careful corrections for divergence of the duct may provide a means of improving recovery factor.

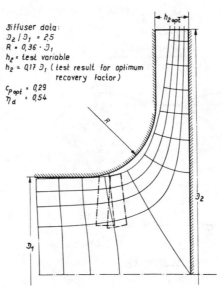

Diffuser data:
$D_2 / D_1 = 2.5$
$R = 0.36 \cdot D_1$
h_2 = test variable
$h_2 = 0.17\, D_1$ (test result for optimum recovery factor)

$c_{p\,opt} = 0.29$
$\eta_d = 0.54$

FIGURE 7.3.4 Comparison of radial (non annular) diffuser with inviscid free cylindrical jet flow impinging wall rectangularly

The most comprehensive survey of diffusers and exhaust ducts of turbomachines is probably the publication of Dejc & Zaryankin [7.4]. They present a large number of test results which supplement other contributions mainly in respect of systematic variation of diffuser and exhaust contours, the dependence on swirl angle and the influence of the arrangement of ribs in the diffuser and exhaust duc┤

Up to a definite limit swirling flow at the inlet of a diffuser may influence performance favourably depending upon the diffuser design. Figure 7.3.5 shows the theoretical meridional flow boundaries of a typical steam turbine last stage, which would coincide with this flow path if the flow leaving the last stage is not deflected. A considerable swirl obviously favours an adhering boundary layer at the

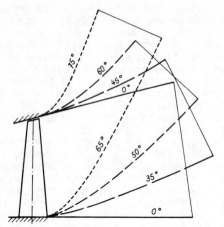

FIGURE 7.3.5 Meridional
contours of straight stream
lines of swirling flow behind a last stage
(hyperbolas)

figures on meridional stream lines are blade exit swirl angles blade
hub / tip ratio : 0.5

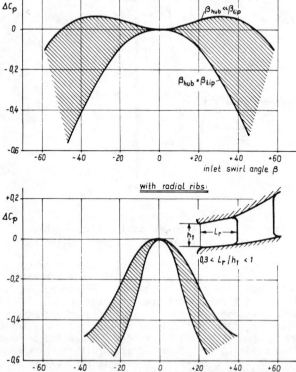

FIGURE 7.3.6 Dependence of
recovery factor upon inlet
swirl angle

outer meridional contour but it may have a detrimental effect on recovery particularly at the inner side of an annular diffuser. It is therefore evident that a
solid body swirl flow behaves more favourably than a free vortex. Also the contours corresponding to the flow angles at part load conditions do not differ very
much from the actual duct contours. This may explain the fairly good recovery
behaviour of last stage diffusers at low volume flows.

A range of test results is shown in figure 7.3.6 as given in references
[7.4,7.6]. The strong detrimental influence of swirl on recovery is noticeable for
arrangements with radial ribs. This will be verified in the next section by an
example in another publication of an exhaust hood test on a commercial design.

7.4 APPLICATION OF LP TURBINE EXHAUST DESIGN

Before discussing commerical designs we will glance at some practical considerations (Fig. 7.4.1). As LP last stages are working with a range of volume flows,
depending upon load or seasonal cooling water temperatures, the chart has taken

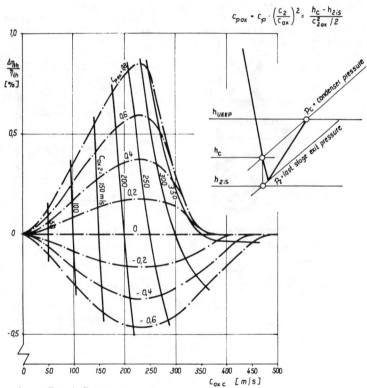

$$c_{pax} = c_p \cdot \left(\frac{c_2}{c_{ax}}\right)^2 = \frac{h_c - h_{2is}}{c_{2ax}^2 / 2}$$

$\Delta\eta_{th}$ = thermal efficiency
c_{pax} = recovery factor, based on $c_{2ax}^2/2$
c_p = recovery factor, based on $c_2^2/2$
c_{2ax} = axial blade exit velocity, based on specific volume behind blade (v_2)
c_{axc} = fictional blade exit velocity, based on specific volume in condenser (v_c)
UEEP = used energy end point

Calculation was made for 8m² last stage annulus area, pitch diameter 2625mm, 3000 rpm,
blade exit area / annulus area = 0,421, η_{th} = 0,40, sonic velocity = 380 m/s

FIGURE 7.4.1 Effect of last stage diffuser
recovery factor on plant thermal efficiency

into account a range of flow conditions for a particular, but representative, final stage. Furthermore, since the static pressure and absolute velocity at the blade exit are not known, particularly under part load swirling flow conditions, the pressure recovery is not referred to the absolute stage leaving kinetic energy, but to the kinetic energy of its (fictional) axial component on the basis of the condenser volume flow as noted in the diagram. A recovery factory $C_{p,ax}$ defined on this basis may well exceed unity at very small volume flows, i.e., large circumferential components.

LP outlet ducts without any guide devices are fairly uncommon today; their recovery factors are about -0.4. Outlet ducts with suitable guide devices may have recovery factors $C_p \approx C_{p,ax} = +0.3$ for the size of customary design exit velocities of 200-300 m/s, which means that there is a rather high potential for improvement. It must also be born in mind that these performance gains can be some 1.6 greater for nuclear units with proportionally greater exhaust flow losses. For axial exit Mach numbers above 0.8, the influence of recovery decreases as the final stage starts to become chocked. High pressure recovery for axial velocities close to sonic velocity is, therefore, of minor interest.

Recent publications report heat rate improvements compared with conventional exhaust hoods of the order of 0.5% [7.7,7.8] from the use of sophisticated steam guides (Fig. 7.4.2). Nevertheless it is notable that a considerable loss is still contributed by reinforcing members [7.8].

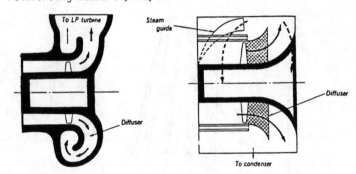

"turbine heat rate has been improved by 0,5% compared with conventional hoods"

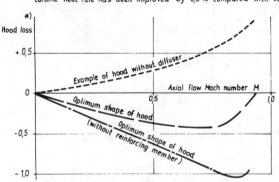

Laboratory test result of hood loss

"increase of thermal efficiency from using a diffuser is about 0,5 percent when it is used in a fossil fuel power turbine at 0,7 - 0,8 axial Mach number"

*) dimension undefined

FIGURE 7.4.2 Seome recent publications
on radial steam turbune diffusers

Some details of a LP last stage diffuser arrangement are given in figure 7.4.3, based on the appropriate performance chart by figure 7.4.1. The local area ratio of the diffuser versus meridional distance is close to unity in the region of the bend, rising only moderately to the exit of the diffuser passage. The recovery factor C_p is low at the higher exit velocities but maintains a value of around 0.25 down to half design volume flow which may be mainly because there are no re-inforcing members or radial ribs.

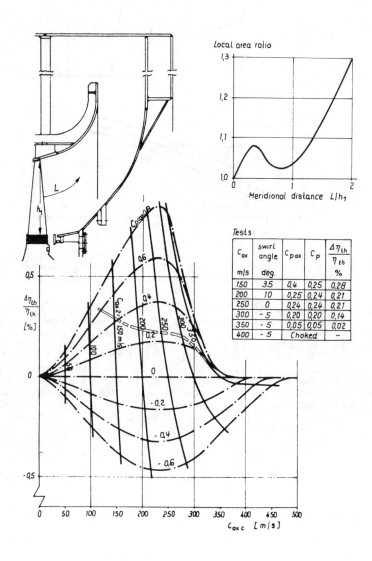

Tests:

C_{ax}	swirl angle	C_{pax}	C_p	$\dfrac{\Delta\eta_{th}}{\eta_{th}}$
m/s	deg.			%
150	35	0,4	0,25	0,28
200	10	0,25	0,24	0,21
250	0	0,24	0,24	0,21
300	-5	0,20	0,20	0,14
350	-5	0,05	0,05	0,02
400	-5	Choked		-

FIGURE 7.4.3 LP last stage diffuser influence on plant thermal efficiency

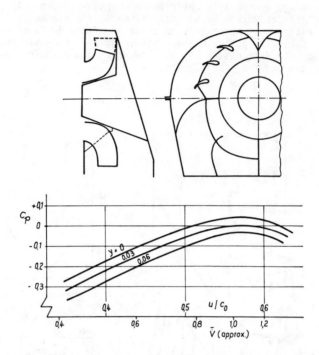

FIGURE 7.4.4 LP-exhaust duct with radial guide vanes – Recovery factor C_p
             versus volume flow ratio $\bar{V}$ and wetness Y, from [7.4]

In [7.4] test results on an exhaust duct with sophisticated arrangement of
vanes and steam guides are reported but the performance seems quite disappointing
particularly at part load (Fig. 7.4.4). An approximate calculation in [7.4] leads
to a similar conclusion, that the radial ribs adversely affect not only recovery
for swirling flows but also at the design point of the last stage.

An interesting survey of possible variations on the Baumann exhaust is given
in [7.9]. The Baumann exhaust (Fig. 7.4.5a) is one way of increasing the discharge
capacity of LP cylinder by utilizing a flow splitter in the penultimate stage, thus
providing an outer annulus area leading directly to the exhaust hood. The so-
called "double-exhaust" (Fig. 7.4.5b) is very similar, featuring equal flows on
both exhausts. The multi-exhaust flow path is characterized by a flow division
in all LP stages (Fig. 7.4.5c). In all cases the blading in the outer exhaust
works on approximately the same velocity ratio (u/C) as the inner blades. The
flow path with division of the flow behind the penultimate stage and deflection
of the upper flow by 180° is reported to be one of the most promising alternatives (Fig. 7.4.5d).

The number of vanes and ribs of these designs nevertheless rises the possibility of detrimental effects on part-load efficiency as well as the threat of
erosion of the bucket trailing edges during no load operation. A study of this
latter problem is given in [7.10]. The no-load flow in a hood is the more complicated the higher the number of separate flow passages adjacent to the blades. An
example of the flow pattern in an exhaust hood is given in figure 7.4.6. Erosion
of the last stage trailing edges is rarely reported on exhausts without ribs.

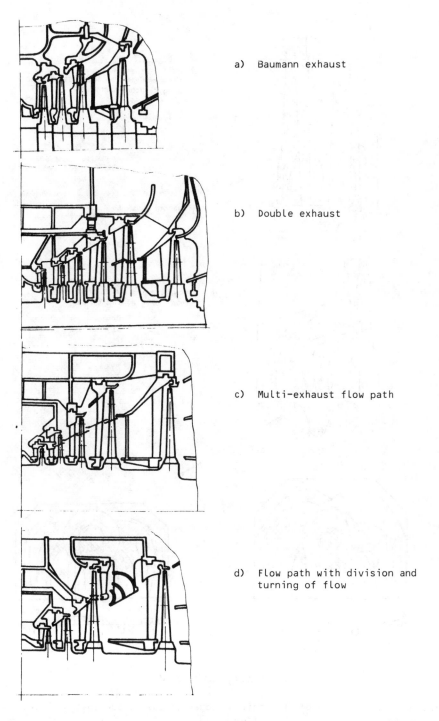

a) Baumann exhaust

b) Double exhaust

c) Multi-exhaust flow path

d) Flow path with division and turning of flow

FIGURE 7.4.5   Flow paths incorporating multi-exhausts from [7.9]

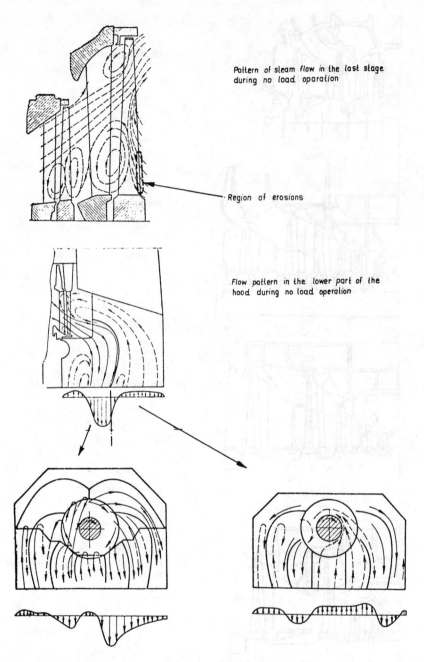

Pattern of steam flow in the last stage
during no load operation

Region of erosions

Flow pattern in the lower part of the
hood during no load operation

Velocity distribution

FIGURE 7.4.6 Erosion of the trailing edges of last stage moving blades,
from [7.10]

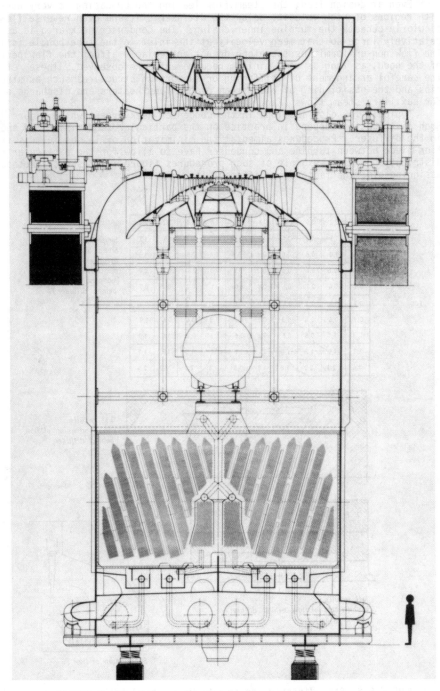

FIGURE 7.4.7 Exhaust hood, bleed and heater arrangement

Even at design load, the steam flow leaving the LP casing is very non-uniform, with regions of high velocity below the diffuser exit and even reverse flow, particularly beneath the turbine inner casing. The condenser neck area is usually relatively large so that mean velocity at the inlet to the tube bundle is low. The flow non-uniformities therefore do not appreciably affect the flow performance of the hood, as long as some minimum requirements are observed. These refer to the careful arrangement of stiffening or supporting structures such as struts or ribs and the positioning of bled-steam pipes, feed heaters and discharge devices for auxiliary steam flows (Fig. 7.4.7).

Typical flow distributions in the condenser neck are shown in figure 7.4.8. Such distributions depend in practice on the particular subdivisions of the flow in the diffuser. Procedures aimed at the minimization of losses in this transition piece between turbine and condenser have to allow for the particular flow distribution. As an example of such procedures figure 7.4.8 shows a method of assigning blockage factors to different flow areas.

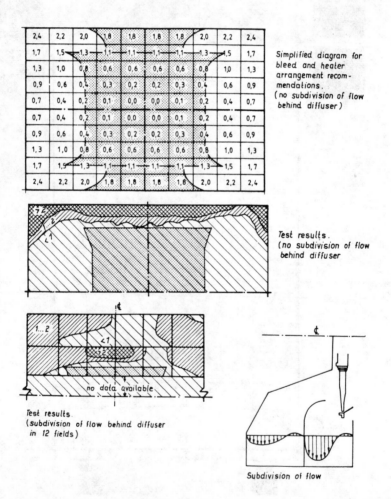

FIGURE 7.4.8 Flow distribution in LP-exhaust hood (figures are velocity/ average velocity ratio)

7.5 CONCLUSION

The designer is at present unlikely to find a reference work presenting a consistent and reasonably complete approach to the optimization of turbine non-rotating passages. Even the otherwise comprehensive treatment of Traupel [7.2] on thermal turbomachinery is confined to a few remarks on the limited possibilities of satisfying aerodynamic requirements and recommends individual testing for the particular application and arrangement. The interpretation of test results will often be difficult as the flow distortions and leakage flows of the actual turbine must be taken into account.

It would be unfair to blame the steam turbine designer for compromising aerodynamic requirements. Selecting a geometry on the basis of the optimization of an isolated parameter nearly always means a deterioration in overall performance. Thus future designs may continue to feature flow passages which are difficult to analyze theoretically and testing of specific geometries will continue to be necessary. However, every endeavour should be made to improve the aerodynamic performance of the exhaust system, the potential for increasing overall plant efficiency being estimated to be about one percent reduction in heat rate.

NOMENCLATURE

A	area
$\bar{A}$	area ratio
B	blockage factor
C	velocity
C_p	recovery factor
D	diameter
h	height or enthalpy
L	length
$\bar{L}$	length ratio L/h
M	Mach number
p	pressure
R,r	radius
Re	Reynolds number
u	peripheral speed
v	specific volume
$\dot{V}$	volume flow
Y	wetness fraction
α	swirl angle
δ	displacement thickness
η_d	diffuser efficiency
η_{th}	thermal efficiency
Δ_p	diffuser pressure rise

Subscripts

ax   axial
C   condenser
is   isentropic
1   diffuser inlet
2   diffuser outlet

REFERENCES

7.1 Sovran, G. & Klomp, E.D.: Experimentally Determined Optimum Geometries for Rectilinear Diffusors with Rectangular, Conical or Annular Cross-Section. In: *"Fluid Mechanics of Internal Flow"*, Amsterdam, *Elsevier*, 1976.
7.2 Traupel, W.: Thermische Turbomaschinen, Bd. 1, Heft 3. Auflage, *Springer* 1977.
7.3 Takchira, A. et al.: An Experimental Study of the Annular Diffusers in Axial-Flow Compressors and Turbines. *Toyko Joint Gas Turbine Conference*, 1977.
7.4 Deich, M.Y. & Zaryankin, A.Y.: Gas Dynamics of Diffusers and Exhaust Ducts of Turbomachines. *Moskau 1970/Foreign Technology Division, U.S. Department of Commerce*, FTD-MT-24-1450-71.
7.5 Albring, W.: Angewandte Strömungslehre. Dresden, *Steinkopf*, 1970.
7.6 Suter, P. & Girsberger, R.: Strömungstechnische Gestaltung des Austrittsstutzens von Axialmaschinen Traupel Festschrift. Zürich, *Juris*, 1974.
7.7 Sohma, A.: Development of Large Steam Turbine with Higher Efficiency. *Hitachi Review*, Vol. 27, No. 3, 1978.
7.8 Hirota, Y.: Recent Technology on Large Steam Turbines. *Mitsubishi Heavy Industries, Ltd., Technical Review*, October 1978.
7.9 Savonov, L.P. & Nishnevich, V.J.: The Ways of Increasing the Discharge Capacity and Engineering Improvement of High Power Steam Turbine Low Pressure Cylinders. *The Institute of Mechanical Engineers, Design Conference 1979 on Steam Turbines for the 1980's*, 1979.
7.10 Lagun, V.P. et al.: Erosion of the Trailing Edges of Last Stage Moving Blades of Steam Turbines. *Teploenergetika*, Vol. 24, No. 10, 1977.
7.11 Lagun, V.P. et al.: Full Scale Tests of the Exhaust of a High-Capacity Steam Turbine. *Teploenergetika*, Vol. 22, No. 2, 1974.

CHAPTER 8

Condensers for Large Turbines

8.1 THERMAL DESIGN

B.J. Davidson

8.1.1 Introduction

The steam condenser on a present day turbine generator unit is a large and comparatively sophisticated heat exchanger. For a 660 MW unit, (Fig. 8.1.1), the condenser contains approximately 400 km of tubing and occupies a space about 500 m^3. It is called upon to create a vacuum equivalent to an absolute pressure of typically 40 mbar and transfer latent heat from steam to cooling water at a rate of about 850 MW. There are nominally 20 000 tubes in a 660 MW unit condenser, arranged horizontally.

In the majority of recent designs, the cooling water is arranged to flow in a single pass through the condenser, in some cases creating a different vacuum in each low pressure turbine cylinder as its temperature rises (multi-vacuum arrangement).

The design condenser vacuum is ultimately determined from the study of plant installation and cycle operation costs. Figure 8.1.2 shows the manner in which thermodynamic efficiency is reduced if the condenser fails to achieve the design vacuum. Conversely, any improvement in vacuum even beyond the design value will increase efficiency, the limit being reached when the steam flow through the final turbine stage becomes choked. In the UK at today's prices, every millibar improvement in the vacuum saves about £280 000 per annum in fuel costs on a 2000 MW station. There is approximately 55 GW of plant in the CEGB so the savings can be significant. In the early 1970's, terminal temperature differences were up to 7°C so that improvements of the order of 4°C (10 mb vacuum) appeared possible. Today, in the UK terminal temperature differences are much closer to the practical limit but scope for improvement still remains.

Apart from rejecting heat at the lowest possible vapour temperature and therefore pressure, a secondary purpose of the condenser is to recover the feedwater for return to the boiler. Since the lowest pressure in the cycle is within the condenser, all the non-condensible gases that either leak into the plant or generate due to feedwater treatment collect there and must be removed. Silver [8.1] in his review of the theory of surface condensers also points out that a condenser should produce condensate with a small amount of subcooling to minimize the need to reheat it back to saturation temperature in the boiler and to minimize absorption of air and other non-condensibles. Such gases must be removed from the condenser, but steam withdrawn at the same time must be kept as small as possible to minimize the need for feedwater make-up into the boiler. The condenser should thus be reheated close to saturation conditions by contact with live steam, and the gases should be cooled just prior to their removal. These requirements should be met whilst keeping the vapour pressure drop as small

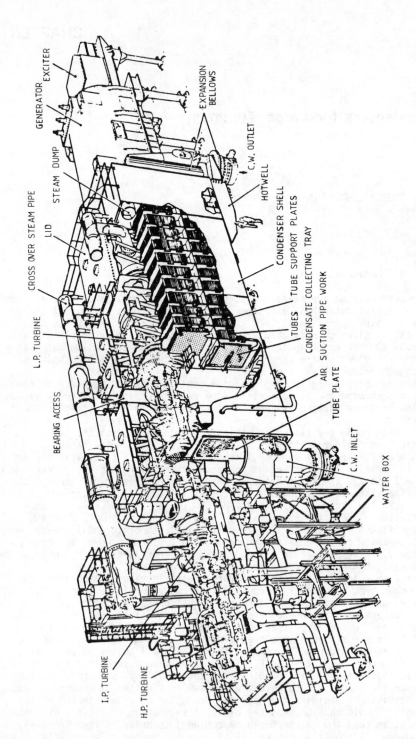

FIGURE 8.1.1   660 MW turbogenerator at Hinkley Point 'B' Power Station

GENERATOR EXCITER

EXPANSION BELLOWS

C.W. OUTLET

CROSS OVER STEAM PIPE

LID   STEAM DUMP

HOTWELL

CONDENSER SHELL

TUBE SUPPORT PLATES

L.P. TURBINE

TUBES

CONDENSATE COLLECTING TRAY

AIR SUCTION PIPE WORK

BEARING ACCESS

TUBE PLATE

C.W. INLET

WATER BOX

I.P. TURBINE

H.P. TURBINE

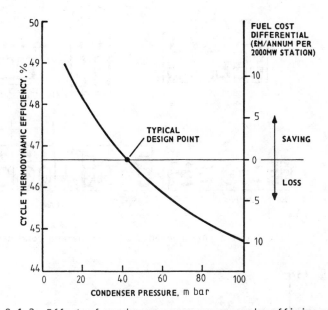

FIGURE 8.1.2 Effect of condenser pressure on cycle efficiency

as possible. Condenser designs have evolved with these points in mind.

8.1.2 Thermal Design Practices

8.1.2.1 *Evolution of Condenser Designs*

Two most important features of a successful condenser tube nest design are :
(1) to minimize the effect of inert gas blanketing and
(2) to minimize the pressure drop.
     In a well designed condenser, the non-condensing gas concentration in the main condensing section, which comprises 90-95% of total surface area, will not in general exceed a few per cent by weight. An air cooling section alone should experience non-condensing gas concentrations greater than this, up to the normal exit value of 25-30% by weight, and should be designed to handle these by ensuring that the velocities are kept high enough to purge non-condensible gases from the condensing surface.
     From these basic design requirements good design principles have evolved :

(a) Live steam must not be allowed to flow to the air cooler without first flowing across the condensing surface. Steam bypassing has the same effect as a high air leakage rate, insufficient suction is produced by the air pumps or ejectors and air pockets develop.
(b) Live steam must not be allowed to envelope the tubes in such a way as to flow radially inwards thereby trapping non-condensing gases in a pocket.
(c) Steam velocity need only be high where the non-condensing gas concentration is high.
(d) Access lanes can be provided to reduce pressure drop when steam velocities are high.

     With these good design principles in mind, it is interesting to briefly reflect on the evolution of condenser designs over the last century since the advent of the steam turbine. For broad details Sebald [8.2] provides an excellent review of the history of surface condensers.

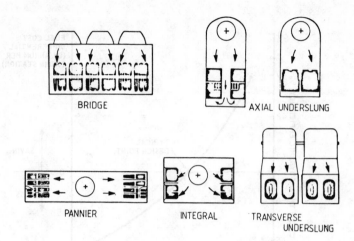

FIGURE 8.1.3   Arrangements of turbine and condenser

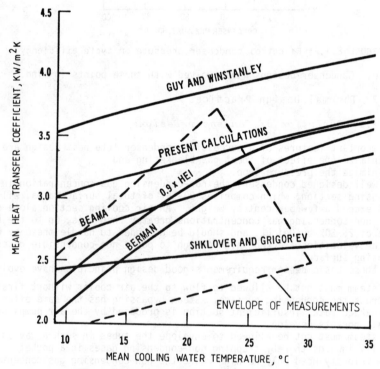

Values adjusted to C.W. velocity = 2.0 m/s admiralty brass tubes.
For the HEI and Berman curves, it is assumed that
$T_{CW\ MEAN} = (T_{CW1} + 5)°C$

FIGURE 8.1.4   Measured range and predicted values of
mean heat transfer coefficient for large condensers

   One important historical factor in the design is the linkage of the condenser to the steam turbine. In the UK most condensing plants in the early 1960's
were placed under the turbine in the underslung arrangement (Fig. 8.1.3). The
condenser tubes were placed horizontally transverse to the turbine shell. The
500 and 660 MW unit condenser designs have evolved from this underslung arrangement as described by Rowe [8.3] and shown in figure 8.1.3.
   The trend towards reducing plant costs by reducing the number of turbine
low pressure cylinders (from 3 to 2 for this unit size) has meant a return to the
transverse underslung condenser arrangement. This has also allowed development
of the modular concept with each turbine cylinder having its own condenser. The
single pass cooling water arrangement has been retained with the condensers pro-
jecting some distance on each side of the turbine.
   Present designs are now based on condenser shells which are more generously
sized so that steam pressure losses are minimized. In addition more thought is
given to the positioning of air vents which ideally should be placed at the lowest pressure points within the tube nests with the air extracted finally at the
coldest point of the condenser. Cold end venting is used to take account of the
different rates of condensation along the tube length due to cooling water temperature variation. Steam which penetrates the nest to the venting duct at the
warm end (cw outlet) of the condenser is transported down the duct and condensed
at the cold end. The residual steam and the condensibles are finally discharged
to the extraction plant at the coldest point of the condenser. This ensures the
lowest possible mixture temperature.

8.1.2.2 Standards for Thermal Rating

Most condensers are thermally rated by choosing a mean heat transfer coefficient,
α, from an empirical formula or standards similar to those proposed by HEI in the
United States [8.4] or BEAMA [8.5]. The surface area required to produce a design vacuum is then derived given values of steam flow, cooling water flow and
temperature. These standards suppose that the overall heat transfer coefficient
is proportional to the square root of the cooling water velocity [8.6].
   Based on early experimental results of Orrok [8.6] and later tests by HEI
the overall coefficient of heat transfer is postulated to be of a form

$$\alpha_0 = C \, F_1 \, F_2 \, F_3 \, \sqrt{v} \tag{8.1.1}$$

where the coefficient C depends upon tube outside diameter and the other coefficients F_1, F_2, F_3 are correction factors for fouling, tube material and wall
thickness, and cooling water inlet temperature respectively. v is the cooling
water velocity.
   In the UK the method of determining the heat transfer coefficient (HTC)
stems from the classic paper of Guy & Winstanley [8.7]. Guy & Winstanley used
a mean steam temperature when deriving their formula. The UK design code
BEAMA (and HEI) uses the more convenient steam inlet temperature and α_0 may be
conveniently expressed as

$$\alpha_0 = 2.15 \, \sqrt{v} \, (0.7586 + 0.013 \, T_v - 0.0001 \, T_v^2) \tag{8.1.2}$$

for surface areas greater than 5574 m^2. In figure 8.1.4 curves with these three
codes are plotted.
   It is possible that a better estimate can be obtained for the overall HTC,
by accounting in an empirical manner, for shell side thermal resistances
between the vapour and the cooling water and also the tube nest geometry. For
example, also plotted on figure 8.1.4 are curves based on the empirical formula
of Berman [8.8] and Shklover & Grigor'ev [8.9] which are much used in the USSR.
The latter method is semi-analytical and includes multiplying factors to account
for steam velocity, presence of air and tube nest shape. This method predicts
lower values of overall HTC than the previous methods and is mainly due to low

predictions of steamside heat transfer coefficients.

Wenzel [8.10] reports wide discrepancies between the predicted values of the overall coefficient and published measured values from commercial condensers. A similar picture can be drawn for UK designs up to the 1970's (Fig. 8.1.4). Current design codes thus only provide a guide to the expected heat transfer coefficient. To ensure that a high heat transfer coefficient is realised in a design, it is necessary to pay careful attention to the tube nest arrangement and shell side aspects. Evidence from recently installed plant in the UK confirms this and the data suggests that the increased attention given to condenser plant design has resulted in performances consistently better than predicted by present design codes, [8.11].

In sections 8.1.3 and 8.1.4 the progress that has been made to accurately define shell side heat transfer and pressure drop is outlined. In sections 8.1.5 and 8.1.6 we follow the development of mathematical models which incorporate detailed shell side calculations and which allow for different tube nest arrangements. The computer codes can be used to assess and optimize condenser performance to realize further improved designs.

8.1.3   Heat Transfer

The starting point for shell side heat transfer calculations is the pioneering work of Nusselt in 1916 of laminar film condensation on a single horizontal tube [8.12]. He idealized the problem by assuming, among other things, a pure quiescent vapour and a uniform tube wall temperature. His analysis produced the well known relationship for the heat transfer coefficient :

$$Nu_N = \alpha_N d_0 / k_c = 0.725 \left[Ga \cdot Pr \cdot K \right]^{1/4} \qquad (8.1.3)$$

Equation (8.1.3) fits well or slightly conservatively with data obtained from experiments which satisfy Nusselt's assumptions. Behaviour in an actual condenser can be somewhat different from the idealized conditions of Nusselt. Marto [8.13] highlights the practical complexities in schematic form (reproduced in Fig. 8.1.5). In reality, the pitching and arrangement of tubes in a bank affects condensate flow and can result in condensate flowing sideways (Fig. 8.1.5b) or in discrete droplets (Fig. 8.1.5c) rather than in a continuous laminar sheet (Fig. 8.1.5a). Large vapour velocities can produce dominant shear forces on the condensate offsetting gravitational forces and which strip the condensate away (Fig. 8.1.5d).

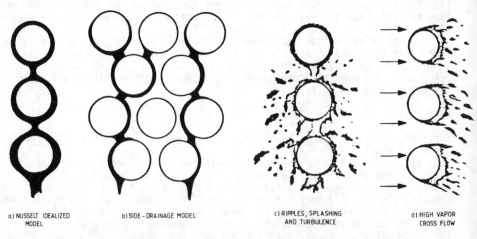

a) NUSSELT IDEALIZED   b) SIDE-DRAINAGE MODEL   c) RIPPLES, SPLASHING   d) HIGH VAPOR
   MODEL   AND TURBULENCE   CROSS FLOW

FIGURE 8.1.5 Schematic representation of condensate flow

Vapour velocity and condensate inundation are significant factors affecting shell side condensation heat transfer and it is demonstrated that they must be accounted for in any accurate condensation analysis.

The most common theoretical treatment is to consider the effects of vapour shear and inundation as separate correction factors applied to the Nusselt equation. Recent studies, outlined in section 8.1.3.4, consider the interaction effects between these physical processes but this work is at an early stage of development. Since computer codes are still generally based on heat transfer routines in which these effects are decoupled, it is prudent to review these studies and compare results where possible with the latest correlations.

The combined average HTC with n tubes in a vertical row is written as

$$\overline{\alpha}_n = \alpha_N \, \phi_I \, \phi_{sh}$$

where ϕ_I is a correction to the single tube Nusselt expression to take into account condensate inundation and ϕ_{sh} is a correction factor for vapour velocity effects. Expressions for ϕ_{sh} and ϕ_I are developed in the following sections.

8.1.3.1 Effect of Vapour Velocity

Theoretical models, based on single tube research, have been proposed from which tube bank correlations have been inferred. Single tube research results are therefore first reviewed as this helps to define the concepts of the physical mechanisms involved.

Single tube analysis

When the vapour surrounding a horizontal tube is moving at a high velocity, the analysis of film condensation is affected in two important ways :

(1) The surface shear between the vapour and the condensate must be included (i.e., the local vapour flow field must be known), and
(2) the effect of vapour separation, its onset and its subsequent influence upon condensate flow, must be accurately treated.

Theoretical investigations of laminar film-wise condensation of flowing vapour have proceeded along two directions. Firstly there are the investigations dealing with refining and developing the Nusselt theory, which are based on solutions of the equation of motion of the condensate film only. These condensate film theories for a flowing vapour relate to the functional form for α/α_N as follows [8.14] :

$$\alpha/\alpha_N = f(II)$$

where $\qquad II = Re_v^2 \,/\, (Ga \cdot Pr \cdot K) = Fr \,/\, (Pr \cdot K).$ $\qquad\qquad$ (8.1.5)

It is assumed that interfacial shear stress during condensation is analogous to a non-condensing gas flowing over a dry surface. In practice, shear stress will be increased and the film thinned, due to momentum transferred to the condensing vapour.

The second type of investigation accounts for the momentum shear stress by basing models on solutions to the equations of a two-phase laminar boundary layer incorporating the interface between liquid and vapour phases. The flow of the bulk vapour is assumed in this case to be potential.

In this more general case, surface shear depends on the ratios of the density and viscosities of vapour and condensate and more complex forms for α/α_N are postulated :

$$\alpha/\alpha_N = f\,(II, S)$$

where
$$S = Pr_c \cdot K \left[\rho_v \mu_v / \rho_c \mu_c \right]^{1/2} \qquad (8.1.6)$$

With condensation on a transversely swept horizontal tube by α and α_N we normally mean tube perimeter average values of the coefficients of heat transfer. In the case of separated vapour flow α will depend on the angle ϕ at which the flow separates from the surface.

Noteworthy studies, based on this two-phase boundary layer approach, are those by Shekriladze [3.15] and Fujii et al. [8.16]. They both obtained expressions which model gravity dominated and vapour shear dominated conditions. Both sets of authors ignored condensate layer pressure, inertia forces, energy convection and condensate subcooling. Such approximations may be justified in some but not in all cases [8.17]. Both analyses were for non separated vertical downflow of vapour and for $0 \leqslant II \leqslant 30$ the difference between predictions is relatively unimportant.

The approximating equation of Fujii is

$$\frac{\alpha}{\alpha_N} = \left[1 + 2.38(1+S)^{4/3} II \right]^{1/4} \qquad (8.1.7)$$

Condensers in modern power stations operate with values of II up to 30 and a typical range for S is 0.2 to 2. Small values of S correspond with high heat fluxes.

In figure 8.1.6, equation 8.1.7 is plotted, together with existing single tube experimental data. A trend of increasing heat transfer ratio can be seen. The scatter in the data is largely only at high values of II. Equation (8.1.7) also correlates the data quite well with S. Equation 8.1.7 should, however, be used with caution to evaluate the extent to which vapour shear increases the heat transfer coefficient since there is an apparent tendency to overpredict. This may in part be due to the assumption that there is no separation of the boundary layer as it flows around the tube and separation can occur somewhere between 82 and 180° from the stagnation point. Where separation occurs, the film rapidly

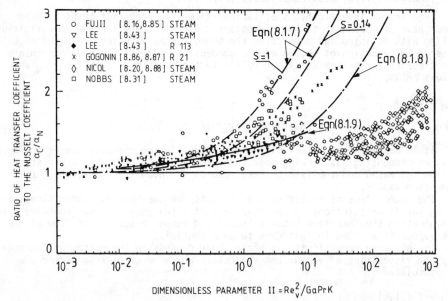

FIGURE 8.1.6 Comparison of single-tube measurements with theory

thickens downstream of the separation point and as a result heat transfer is deteriorated. Recent complete analyses by Fujii and co-workers have considered the effect of variation in, for example, the separation point but surprisingly such refinements give results which differ only slightly from equation 8.1.7. In addition, whilst these equations were obtained for vapour downflow, Honda & Fujii [8.18] have shown that the results for horizontal steam flow are almost identical to those for vapour downflow.

Other studies [8.15] recommend quite conservative interpolation formula to account for the effect of separation. For example, if it is assumed there is no heat transfer beyond the separation point and a minimum angle of 82° is chosen the most conservative equation results. Heat transfer is then reduced by typically 35%. Such an approach is proposed by Lee & Rose [8.19] and their equation is of the form

$$Nu/Re_{TP}^{1/2} = 0.416 \left[1 + \left(1 + \frac{9.47}{\Pi} \right)^{1/2} \right]^{1/2} \tag{8.1.8}$$

Except for the suspect data at very high velocities [8.20], all the data on figure 8.1.6 lie above the prediction of Lee & Rose verifying its conservative nature.

Tube bank analysis

An intermediate step between performing single tube experiments and condensing bank experiments is to examine a single condensing tube in dummy bank. The complicating effects of inundation from surrounding tubes is still avoided. Curves have been fitted to such data and the Berman equation :

$$\alpha/\alpha_N = 1.28 + b\log_{10}\Pi$$

$$b = 0.12 \qquad\qquad 0.01 \leqslant \Pi \leqslant 1 \tag{8.1.9}$$

$$= 0.21 \qquad\qquad 1 \leqslant \Pi \leqslant 15$$

plotted on figure 8.1.6 is one example. The equation fits the vertical downflow experimental data of Berman and Tumanov [8.24]. Predictions from Berman and Tumanov's own equation (see Fig. 8.1.12) and equation (8.1.9) are virtually co-incident over the data range $0.1 < \Pi < 15$. These equations are singled out for comparison since they are consistently used in condenser performance computer codes (in conjunction with other correction factors for inundation, air, etc). The effects of separation are implicitly reflected in equation (8.1.9) and it is interesting that these bank predictions agree reasonably well with those from the conservative single tube interpolation equation (8.1.8) recently proposed by Lee & Rose. The equation of Lee & Rose has the advantage though of satisfying the extremes of gravity and velocity dominated condensate flow.

Finally, it is apparent from figure 8.1.6 that more experimental data are needed for larger values of Π and also smaller values of S (high heat fluxes) in order to verify the dependence upon S.

8.1.3.2 *Inundation in Tube Banks*

In the absence of vapour velocity then as condensate flows by gravity on to lower tubes in the bundle, the condensate should thicken around a tube and the condensate heat transfer coefficient should therefore decrease. Jakob [8.22], Kern [8.23] and Eissenberg [8.24] amongst others, extended the Nusselt analysis for film condensation heat transfer on a vertical row of horizontal tubes. Each author postulated a different condensate flow behaviour (e.g. continuous sheet,

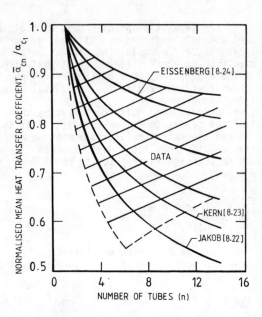

FIGURE 8.1.7 Comparison of experimental data and correlations
showing the effect of condensate rain on heat transfer

droplets, side drainage) and from figure 8.1.7 the data shows considerable scatter around each of these predictions with the Jakob expression predicting the most conservative result.

Kutateladze [8.25] and Fuks [8.26], by a similarity analysis of the condensation equations, derived a non dimensional equation which accounted for the predominant physical mechanism :

$$\overline{\alpha}_n/\alpha_1 = f \ (w/c) \tag{8.1.10}$$

where w = total inundation rate on the nth tube

c = condensation of the nth tube

$\overline{\alpha}_n$ = mean heat transfer coefficient for n tubes

For a relatively closely spaced bank with high vapour velocities Fuks showed that

$$\overline{\alpha}_n/\alpha_1 = [w/c]^{-s} \tag{8.1.11}$$

with $s = 0.07$. Other authors (e.g. Wilson [8.27]) quote higher constant values for s (up to 0.223) with large values of s generally providing a more conservative correction factor for inundation.

8.1.3.3   *Correction Factors to the Nusselt Equation*

Grant & Osment [8.28] proposed the following expression

$$\overline{\alpha}_n/\alpha_N = [w/c]^{-0.223} \left[1 + 0.0095 \ Re_v^{11.8\sqrt{Nu_N}} \right] \tag{8.1.12}$$

which is a combination of a term of the form of equation (8.1.11) for inundation and Berman & Tumanov's [8.21] own equation (analogous to equation (8.1.9)) to account for vapour velocity effects.

A general design method could utilize a slightly more conservative combination of correction factors, namely equation (8.1.8) for the effect of vapour velocity and the Kern method for inundation :

$$\bar{\alpha}_n = n^{-1/6} \cdot \frac{k_c}{d_0} \cdot Re_{TP}^{1/2} \cdot 0.416 \left[1+ \left(1+ \frac{9.47}{11} \right)^{1/2} \right]^{1/2} \tag{8.1.13}$$

8.1.3.4 *Equations for the Combined Effect of Vapour Shear and Inundation*

Two authors, Brickell [8.29] and McNaught [8.30] have attempted to embody interactive effect of a vapour flow and condensate inundation into their correlations.

Brickell recognized that s in equation (8.1.11) cannot be a constant but is perhaps a function of vapour velocity, flow direction, pitch to diameter ratio and tube arrangement as shown in figure 8.1.8. This shows that the trend is towards a low exponent only at high vapour velocities. Also upflow geometries require higher exponents than downflow geometries and wider pitches require higher exponents for both up- and downflow.

Brickell postulates that the condensate should be treated as having two components, a film flow and a drop flow with droplet entrainment occurring due to the local vapour velocity conditions.

To account for the interaction between vapour velocity and condensate inundation, Brickell used the published vapour shear correlation of Berman, (Eqn. 8.1.9) and an inundation correlation is derived to match measured heat transfer data. The final equation (obtained for pressures between 23 and 157 mb) has the form :

$$\alpha_n/\alpha_N = N^{-s} [a+b \log_{10} II] \tag{8.1.14}$$

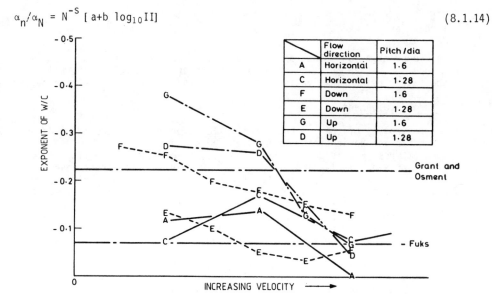

FIGURE 8.1.8 Effect of vapour velocity, flow direction and tube spacing on inundation exponent [8.29]

where N is no longer the ratio w/c but relates to the amount of condensate actually resident on the tube. The expression for N accounts for droplet entrainment in the steam flow. N is therefore a function of vapour velocity and is correlated against Froude number.

The procedure for determining s is described in [8.11] with s being dependent upon the principle flow directions.

Comparing predictions from equation (8.1.12) with the downflow data used to derive equation (8.1.14) there is a tendency for equation (8.1.12) to overpredict at low heat transfer levels where inundation is present and underpredict some of the highest heat transfer coefficients which were measured with little inundation.

McNaught [8.30] suggests that shellside condensation at high vapour velocities might be treated as two-phase forced convection. He assumes that high velocity data may be correlated by an expression of the form

$$\alpha_{sh}/\alpha_L = a X_{tt}^{-b}$$
                                                                                           (8.1.15)

where α_L is the liquid phase forced convection coefficient across a bank of tubes and X_{tt} is the Lockhart-Martinelli parameter

$$X_{tt} = \left[\left(\frac{1-x}{x}\right)^{1.8} \left(\frac{\rho_v}{\rho_c}\right) \left(\frac{\mu_c}{\mu_v}\right)^{0.2} \right]^{1/2}$$
                                                                                           (8.1.16)

The coefficients a and b are empirical constants which for the data of Nobbs [8.31] for downflow of steam at atmospheric pressure in staggered and in-line tube bundles takes the values 1.26 and 0.78 respectively.

For gravity controlled condensation he uses equation (8.1.11) for α_{gr} with s = 0.22 for an in-line bundle and 0.13 for a staggered bundle. He adds each contribution to obtain

$$\bar{\alpha}_n = \left[\alpha_{sh}^2 + \alpha_{gr}^2 \right]^{1/2}$$
                                                                                           (8.1.17)

McNaught found that 90% of Nobbs' data could be predicted to within ± 25% using this method. Equation (8.1.17) shows better agreement than equation (8.1.13) with the latter being conservative with respect to most of the data as expected.

Future development

The generality of the two phase multiplier approach for other flow directions in sub-atmospheric pressures needs to be demonstrated.

None of these models so far presented account of the possibility of movement of condensate axially along the tube. A number of researchers have observed such movement [8.32] and the effect of inundation can be diminished in this way if conditions exist such that the flow of condensate is confined to a part of the tube bank. Fujii [8.33] proposes two theoretical models for the spreading of condensate axially and if such flow behaviour can be realized with certain reproducibility better theoretical predictions may be possible by developing his ideas. Indeed, Shklover and Buevich [8.32] propose the sloping of condenser tubes to promote drainage via the tube support plates.

Two authors, Berman [8.34] and Fujii [8.35] provide comprehensive compilations of film condensation data on bundles of horizontal tubes. The variation in heat transfer performance data is substantial and is probably more influenced by design and operational factors rather than the procedures for processing the experimental results. For example, Fujii has performed experiments with steam

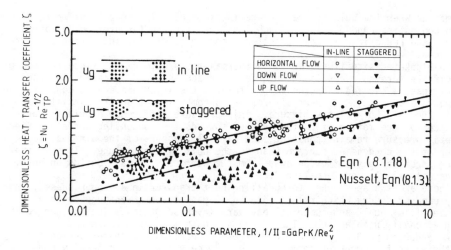

FIGURE 8.1.9 Relationship between dimensionless heat transfer coefficient and II for horizontal downward and upward flow in in-line and staggered banks [8.35]

flowing either downward, horizontally or upward in a fully condensing bank. He correlates his horizontal and vertical flow data by an equation

$$Nu/Re_{TP}^{1/2} = 0.96\, II^{-0.2} \tag{8.1.18}$$

By inspection, this equation is correlated in terms of single tube parameters with no explicit dependence on row number or tube bank geometry.

Equation (8.1.18) is plotted on figure 8.1.9 together with the experimental data, obtained for both in-line and staggered tube bundles. It is significant that there is very little difference between downward and horizontal flow data. All such data lie above predictions using Nusselt's equation. However, upward flow data are as much as 50% lower for values of II in the range $2 < II < 10$ and some are less than predictions using the single tube, zero velocity expression (8.1.3).

The considerable influence of steam velocity in Fujii's experiments for horizontal and vertical flow is illustrated in figure 8.1.9 since all experimental values of α/α_N exceed unity. This is clearly not the case for the upflow data. Here, condensate hold-up occurring locally round a tube due to hydrodynamic conditions, could be responsible for thicker condensate films on the tubes and measured deterioration in performance. Condensate hold-up may also be responsible for poor heat transfer measurements in some horizontal flow experiments [8.36]. Alternatively Eissenberg argues that condensate carried horizontally or diagonally across the tube rows largely contributes to the observed increases in steamside heat transfer coefficients with increased vapour flow, since the level of condensate inundation is effectively decreased. Both interpretations have credence depending on local conditions and tube bank pitch to diameter ratios.

The results of Fujii and the compilation by Berman highlight the need to perform more experiments with test conditions and geometries comparable with operating plant.

8.1.3.5 *Effect of Non-Condensible Gases*

It is well known that in the presence of a small amount of non-condensible gas, condensation heat transfer rates can be significantly reduced. Reductions are

most marked when the bulk of the mixture is stagnant. The heat transfer rate in
forced convection condensation is much less sensitive to the presence of air
although the reductions in heat transfer become more significant as the pressure
is reduced.

In blanketed situations, an added thermal resistance exists since the vapour
must diffuse through a gas layer in order to reach the cold surface. An incondensible gas is carried with the vapour towards the interface and as condensation
of vapour takes place, its partial pressure becomes less than its partial pressure in the main bulk of the mixture. This pressure difference provides a driving force for the motion of further vapour towards the condensing surface. Since
the total pressure remains constant, the partial pressure of the air must be
greater at the condensing surface than in the bulk of the mixture. This is the
driving force for gas diffusion away from the surface. The air builds up until
motions are exactly counterbalanced.

There are two approaches to modelling heat transfer in a vapour phase : the
widely used procedure which relies on a "heat-mass transfer analogy" with a
"stagnant film model" and numerical boundary layer solutions which are generally
not in a readily usable form. The former approach was first expressed by
Colburn & Hougen [8.37]. The basic weakness of heat-mass transfer analogy is
that the velocity component normal to the surface v_0, say, is generally non-zero
in the mass transfer case while it is generally zero in the heat transfer case
from which a solution to the mass transfer problem is inferred. The weakness is
recognized by some authors (Berman [8.32]) who apply compensating correction
factors.

If the surface is assumed porous and transpiration suction is imposed so
that the normal velocity component is the same in the heat transfer case as in
the mass transfer case, then the two problems are strictly identical. Rose [8.39]
pursues this approach since (a) numerical solutions of the flat plate heat transfer problem have been obtained [8.40] for the case where $v_{0\infty}x^{-1/2}$ (where x is
distance measured along the surface) and fortunately (b) the thickness of the
condensate layer for condensation in the presence of non-condensible gas leads
to the same dependence of v_0 on x.

The general form for the Sherwood number Sh in the case of a condensing
cylinder in a steam-air mixture is approximated by Rose as follows :

$$\text{Sh Re}_v^{-1/2} = 0.57 \text{ Sc}^{1/3}(1+\beta. \text{ Sc})^{-1} + \beta \text{ Sc} \qquad (8.1.19)$$

where

$$\beta = - \frac{v_0}{u_\infty} \text{ Re}_v^{1/2}$$

By noting that the heat flux Q can be written as :

$$Q = L\rho \frac{D}{d_0} (1-\omega) \cdot \text{Sh}$$

where ω is the ratio of the free stream air mass fraction to the condensate
surface air mass fraction, then equation (8.1.19) can be rearranged to give :

$$Q = \frac{L}{2d_0} D_p \text{ Re}_v^{1/2} \left(\frac{\Delta P}{\Delta T}\right)_{sat}^{2/3} \Pi^{-2/3} P^{1/3} \left[1+2.28\text{Sc}^{1/3}(1/\omega-1)^{1/2}-1\right] (T_v-T_{cs})^{2/3}$$

$$(8.1.20)$$

$$\left(\frac{\Delta P}{\Delta T}\right)_{sat} = \frac{II \ P}{(T_v - T_{cs})}$$

where the coefficients of diffusion D and D_p are related as follows

$$D_p = \frac{M \ D}{R_0 T_v} = \frac{D}{R_s T_v}$$

In figure 8.1.10, data [8.42] are compared with equation (8.1.20) and the agreement is good. The Berman & Fuks correlation [8.42] is often sited in computer codes and agreement with equation (8.1.20) is poor for high values of the bulk air-steam ratio where the correlation significantly overpredicts the vapour heat transfer coefficient. Also the correlation behaves incorrectly as $1/\omega$ approaches unity suggesting that the data did not extend to these conditions.

8.1.3.6 Tube Side Heat Transfer

Numerous correlations are available for predicting turbulent forced convective heat transfer coefficients for water flowing inside smooth tubes. A series of experiments were undertaken [8.44] to assess the accuracy of these existing correlations. The popular Dittus-Boelter equation [8.45] was thought to be too conservative [8.46] for design calculations and there is experimental evidence [8.44] to support the Petukhov-Popov correlation

$$Nu = \frac{(\epsilon/8) \ Re \ Pr}{K_1 + K_2 (\epsilon/8)^{1/2} \ (Pr^{2/3} - 1)} \qquad \begin{array}{l} 0.5 < Pr < 2000 \\ 10^4 < Re < 5 \times 10^6 \end{array} \qquad (8.1.22)$$

where

$$\epsilon = (1.82 \ \log_{10} Re - 1.64)^{-2}$$
$$K_1 = 1 + 3.4 \ \epsilon$$
$$K_2 = 11.7 + 1.8 \ Pr^{-1/3}$$

in preference to Dittus-Boelter for geometries and conditions of power condenser operation. On figure 8.1.11 are plotted the two Nusselt (Nu) equations as a function of Reynolds number. The differences between the predictions of Nu vary between 8% and 14% depending on Pr and Re but experimental support for Petukhov-Popov is only available for Re up to 3.5×10^4 with Pr = 6 and 11.6. The classic Eagle & Ferguson correlation [8.47] was found to give reasonable agreement with (8.1.22) at Pr = 6.0 but tended to be too conservative (low) at higher Prandtl numbers. The Sieder-Tate correlation [8.48] predicts values which correspond approximately with those obtained from Dittus-Boelter but the Sleicher-Rouse correlation [8.49] agrees well with predictions from (8.1.22). The Pekukhov-Popov expression has been singled out in preference to Sleicher-Rouse as the former has also accurately predicted a spectrum of constant-property data with fluids other than water.

Current condenser models have tended to use the Dittus-Boelter correlation and further experimental verification to support or otherwise the Petukhov-Popov equation is desirable to reduce, if necessary, design margins and to clarify steam side effects from experiments.

8.1.3.7 Evaluation of the Overall Heat Transfer Coefficient

Including all the thermal resistances between the vapour and the cooling water, the overall heat transfer coefficient α_0 referred to the outside tube area may

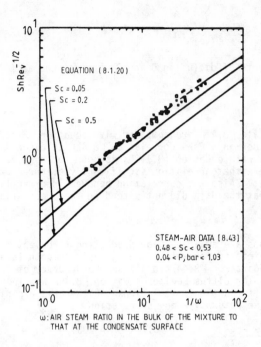

ω:AIR STEAM RATIO IN THE BULK OF THE MIXTURE TO
THAT AT THE CONDENSATE SURFACE

FIGURE 8.1.10 Comparison of predictions from equation (8.1.20) with data
for condensation of steam on horizontal tubes in the presence of air [8.41]

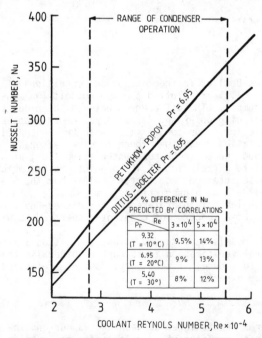

FIGURE 8.1.11 Comparison of waterside heat transfer coefficient

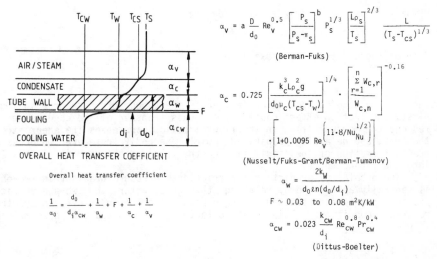

FIGURE 8.1.12 A combination of heat transfer correlations for the evaluation of the overall heat transfer coefficient

be expressed as

$$\frac{1}{\alpha_0} = \frac{d_0}{d_i \alpha_{cw}} + \frac{\ell n\left(\dfrac{d_0}{d_i}\right) d_0}{2 \, k_w} + F + \frac{1}{\alpha_c} + \frac{1}{\alpha_v} \qquad (8.1.23)$$

where k_w is the thermal conductivity of the wall and d_i and d_0 the inside and outside diameters of the tubes respectively. F is a fouling factor. Figure 8.1.12 presents one particular combination of correlations for substitution in equation (8.1.23). The condensate heat transfer coefficient α_c is based on separate correction factors to the Nusselt equation of the top tube.

Only when the above equation is utilized locally, will it be possible to perform more accurate analyses of condensers and improve upon their performance.

8.1.4 Pressure Drop Across Tube Banks

Two phase prediction methods are again necessarily highly empirical and must also be based on very scarce data available in the literature. No published correlation is general enough to reliably account for as wide a range of conditions as exists in even the present available data. A reason for the lack of generality is the attempt to fit a given set of data as precisely as possible by an empirical curve without sufficient regard to possible mechanisms that could affect the form of the function. In an attempt to return to more fundamental bases the Martinelli separated flow approach, which has repeatedly proved successful for tubeside flow has been reinvestigated [8.37]. It is concluded that the original Martinelli model can be adapted to give the best general agreement with the data. It is further recognized, that the basic Martinelli approach can be significantly improved if used in conjunction with flow regime parameters [8.56]. The models are described. Existing correlations are also summarized which pertain almost entirely to adiabatic flow and which can be used if modified to include additional parameters.

In simplistic terms, the static pressure difference between two points in a tube nest can be considered as the sum of the terms :

$$dP/dz = \underset{frictional}{dP_f/dz} + \underset{gravity}{dP_g/dz} + \underset{momentum\ change}{dP_a/dz} \qquad (8.1.24)$$

Each of these terms is now examined.

8.1.4.1 Frictional Term

The occurrence of condensation influences the vapour pressure drop in two ways. First, the suction effect caused by the mass transfer of the condensing vapour reduces the momentum flow in the flow direction, increases the shear stress on the tube, and delays the separation point. Secondly there are two-phase effects due to formation of the liquid film on the tube and entrainment of this liquid into the vapour space.

The problem of predicting the frictional pressure gradient in two-phase flow is generally approached by relating, through a factor, the two-phase frictional pressure gradient to that in a single phase, flowing under certain hypothetical reference conditions. This relating factor is called the two-phase multiplier, ϕ^2 where

$$dP_f = \phi^2_{GO} \cdot \left(dP_f\right)_{GO} \qquad (8.1.25)$$

The subscript GO refers to the vapour flowing at the same mass flow rate as the complete two-phase mixture.

The frictional pressure drop for the vapour-phase is mostly expressed in the form

$$\left(dP_f\right)_{GO} = 1/2\ \xi\ N\ \rho_v\ u^2 \qquad (8.1.26)$$

where ξ is a friction factor, N is the number of restrictions encountered and $\rho_v u^2/2$ is the kinetic head through the restriction based on the average maximum steam velocity in the bundle.

For one dimensional flow over a bundle of standard layout, single phase friction factors ξ may be obtained from a variety of sources, [8.50] and figure 8.1.13 is one example.

In an examination of the available data in the laminar and transition region, an analogy with non-linear porous media can be made which leads to a particularly simple form of correlation for two dimensional flow modelling studies [8.51]. The angular symmetry of bundles leads to constraints in the form of the loss model which renders it isotropic in the limit of zero velocity and gives a general form for the loss relation valid for all flow directions and magnitudes. Butterworth [8.52] and ESDU [8.53] propose correlations but there is little difference in the respective predicting capabilities over the available data range. The concept of isotropy implies that there is no assumed difference in pressure gradient for triangle and rotated triangular (and similarly square) arrays.

Writing

$$\left(\frac{dP_f}{dz}\right)_{GO} = -\frac{C}{d_0}\frac{\rho_v\ U^2}{2} \qquad (8.1.27)$$

C is obtained from values calculated for low flow C_L and high flow C_H with a final value of C obtained from

$$C = \left[C_L^2 + C_H^2\right]^{1/2}$$

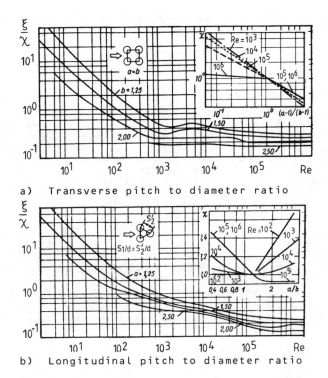

a)   Transverse pitch to diameter ratio

b)   Longitudinal pitch to diameter ratio

FIGURE 8.1.13   Dry pressure loss coefficient for in-line and staggered banks [8.50]

(C_L, C_H are defined in [8.53]). U is a superficial vapour velocity, defined as the fluid velocity in the absence of the tube bundle. These correlations give a good representation of available data in the laminar and transition regions and for triangular bundles a reasonable extrapolation into the fully turbulent regime. There are, however, discrepancies between experimental results from different workers, particularly for rotated square arrays, so there is a need for further experiment.

In order to determine the multiplier, ϕ^2, Brickell [8.29] and Ishihara et al. [8.54] re-examined the Martinelli separated flow approach which assumes that both the liquid and vapour phases experience the same pressure drop but do not have the same velocity. The latter authors tried various forms of pressure drop correlations as a function of the Martinelli parameter, X_{tt} given by equation (8.1.16), and proposed the following general forms :

$$\phi^2_{LO} = \frac{\Delta P_{TP}}{\Delta P_{LO}} = 1 + \frac{C'}{X_{tt}} + \frac{1}{X^2_{tt}} \qquad (8.1.28)$$

for $Re_L > 2000$ and

$$\phi^2_{GO} = \frac{\Delta P_{TP}}{\Delta P_{GO}} = 1 + C' X_{tt} + X^2_{tt} \qquad (8.1.29)$$

for $Re_L < 2000$.

They point out that C' is an empirical adjustment factor which in general is a function of the pertinent two-phase flow variables :

$$C' = f \left[X_{tt}, \eta, \frac{1-x}{x} \right] \tag{8.1.30}$$

The variable η is proportional to the ratio of the vapour shear force to the gravity force, and is a function of the Wallace parameter and the tube spacing to tube diameter ratio. $(1-x)/x$ is the ratio of liquid to vapour present.

The form of the function C' can be determined empirically by catagorizing data according to postulated flow models, e.g. the transition from a gravity dominated regime (stratified, stratified-spray and slug) to a shear dominated regime (spray and annular). Correlations for each regime are determined by drawing a similarity between shell side and tube side behaviour to define transition regions. Such a comparison has been made [8.55] for vertical flow but the final form for C' is still under development.

Ishihara et al. [8.54] use a constant value of C' (= 8.0) to give reasonable agreement principally with air-water flow data for small values of X_{tt} (<0.2). Data are based on small bundles and results reflect a shear controlled regime under low liquid loading. Hopkins et al. [8.11] also propose a functional form for ϕ^2 similar to equation (8.1.29) based on condensing steam tube bank data but the exact form for ϕ^2 has not been published.

Further work is needed on two-phase prediction methods and many authors simply use the drag coefficient for a dry (non-condensing) bank as defined by Zhukauskas [8.50]. The possible consequences of this simplifying assumption are highlighted in section 8.1.4.3.

8.1.4.2 *Gravitational Term*

This term takes the form

$$dP_g = \pm \, g \, \sin\theta \left[r\rho_v + (1-r)\rho_c \right] dz \tag{8.1.31}$$

where θ is the angle of the direction of flow from vertically upwards and r is the void fraction (ratio of cross-sectional area occupied by vapour to total cross-section area).

Significant differences in pressure drop can occur with either upflow or downflow compared with horizontal flow and the differences are due to the two-phase gravitational term which is important when a significant amount of condensate occurs within the bundle. Unpublished air water friction factor data [8.57] show the trends (Fig. 8.1.14).

Monks et al. [8.57] report on a gravitational model based on bank data with condensing steam. They correlate the void fraction against the Froude number but specific details are not given. For downflow the void fraction may be expected to decrease with increasing Froude number (and thus steam velocity) and also to increase with decreasing dryness.

For upflow, flooding may occur when the upward steam velocity is sufficient to prevent condensate draining downwards due to gravity. Condensate flow reversal may occur when the steam velocity is great enough to blow the condensate upwards through the bundle. Monks et al. demonstrate that the gravitational term can be very significant and can be as much as 50% of the frictional term and so for downflow tests for example, measured pressure drop may only be one half the single phase prediction.

8.1.4.3 *Momentum Terms*

Another complicating feature of vapour flow with condensation is that the mass extraction causes the vapour velocity to change throughout the bundle. The

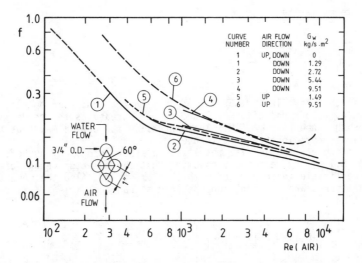

FIGURE 8.1.14 Air-water friction factor data for different flow directions and mass velocity [8.57]

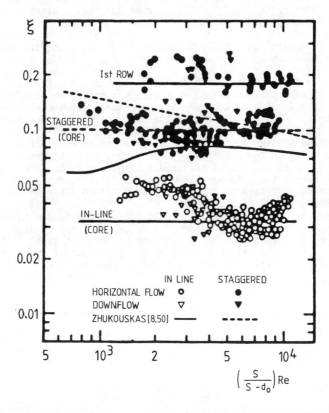

FIGURE 8.1.15 Experimental results for the local drag coefficient [8.33]

momentum terms may be written as :

$$\frac{dP_a}{dz} = \frac{d}{dz}\left[\dot{m}^2\left(\frac{(1-x)^2}{\rho_c(1-r)} + \frac{x^2}{\rho_v r}\right)\right] \qquad (8.1.32)$$

where $\dot{m}$ is the mass flux [8.55]. Depending on the condenser operating pressure, for example, vapour velocity may decrease or increase as the condensing steam traverses the tube rows [8.13]. The expression for dP_a/dz may thus be positive or negative (providing pressure recovery).

To interpret experimental results of pressure drop many authors simplify equation (8.1.32) by reasonably assuming that the momentum change of the condensate is negligible in comparison with that of steam. Also the momentum and frictional terms are "lumped" together and a drag coefficient ξ is defined by the equation

$$\xi = \frac{\bar{\rho}}{2n\ \bar{\dot{m}}^2_{max}}\left[(P_i - P_j) + \left(\frac{\dot{m}^2}{\rho}\right)_i - \left(\frac{\dot{m}^2}{\rho}\right)_j\right] \qquad (8.1.33)$$

This equation is derived for a control volume bound by the ith and j^{th} widest cross-sections and it is assumed that variation in pressure in the cross-sections are small compared with $(P_i - P_j)$. The overbar on ρ and $\dot{m}$ denotes average values between a value for i and that for j.

Both Nicol et al. [8.58] and Fujii [8.33] obtained some recent experimental data for condensation rates and pressure drops during crossflow of steam in small tube bundles. Data were taken with both in-line and staggered geometry and they obtained values for drag coefficient ξ. Fujii's data for horizontal and downflow is presented in figure 8.1.15 with ξ plotted versus steam Reynolds number. The data has a large dispersion but it is clear that the value of ξ for the first row is high and that there is a characteristic difference between the in-line and staggered banks. Also, compared with the drag coefficient for a dry (non-condensing) bank, ξ_D defined by Zhukauskas [8.50], the drag coefficient with condensation is less and substantially so with the in-line bank. This decrease is attributed to the delay in the onset of separation caused by the condensation process. The data of Nicol et al. is not included in figure 8.1.15. Their data exhibited the same trends but the data were lower than Fujii presumably due to the fact that Nicol condensed steam at pressures from 0.2 to 0.8 bar while Fujii condensed steam at pressures 0.01 to 0.07 bar.

Table 8.1.1 summarizes the Fujii data for ξ normalized with respect to the dry ξ_D. For comparison the corresponding values obtained by Nicol et al. and Lee [8.59] are also tabulated. The tests by Lee consist of a simulation of condensation with the suction of air through seven rows of porous tubes. The results agree qualitatively for the case of a staggered bank.

Authors	Tube Arrangement	d_0 (mm)	s/d_0 (trans./ longi.)	1st Row	Core Part	
					Horizontal	Downflow
Fujii et al. [8.89] (condensation at Re = 10^4)	in-line	14.0	1.57/1.57	2.3ξ_D	0.4ξ_D	0.35ξ_D
	staggered	14.0	1.57/1.57	1.8ξ_D	ξ_D	0.85ξ_D
Nicol et al. [8.90] (condensation at Re = 10^4)	in-line	9.525	1.736/1.736	1.4ξ_D	0.27ξ_D	-
	staggered	9.525	1.58/1.80	-	0.75ξ_D	-
Lee [8.59] (suction at Re = 2.5×10^4)	in-line	19.0	1.25/1.25	-	ξ_D or more	
	staggered	19.0	1.25/1.083	-	0.9 ξ_D	
	staggered	19.0	1.25/0.935	-	(0.85-0.9)ξ_D	

TABLE 8.1.1 Drag coefficients in condensation of steam and suction of air against those without condensation and suction (after Fujii [8.33])

It is interesting to note from Table 8.1.1 that for horizontal flow in staggered tube banks, ξ approximately equals ξ_D. Further support for this is provided by Rowe et al. [8.60].

To summarize, pressure drops for downflow and horizontal flow in staggered tube arrangements can be predicted fairly accurately for small tube banks. For large staggered banks, with significant water loading, the work of Brickell et al. suggests that for downflow, values of ξ/ξ_D may be much less than 0.85 (as shown in Table 8.1.1). For upflow the process is complex and the scatter of experimental data is large. Models which identify the various flow regimes are needed for accurate predictions. Such models are at an early stage of development. Finally, experimental data is insufficient to estimate pressure drops for in-line banks. It appears that in-line tube arrangements are effective in reducing drag coefficient and further work should concentrate on detailed measurements of the effect of tube arrangement and pitching on condenser performance.

8.1.4.4 *The Role of Access Lanes to Reduce Pressure Drop*

In order to reduce pressure drop caused by vapour passing through the first rows of tubes at high velocity it is common practice to provide access lanes, i.e., to omit tubes to form lanes for vapour to penetrate the bundle without being forced through numerous rows of tubes. In providing access lanes care must be taken to ensure that the velocity of the vapour among the tubes bypassed by the access lanes remain sufficiently high to prevent the formation of local air pockets.

Small scale tests have been published [8.61], [8.63] but data is limited. Bell [8.62] covers the experimental work for single phase flow and offers a correlation. Lee et al. [8.63] compare the predictions of Bell with their own data obtained using air flowing over porous tubes with and without suction. Their preliminary data for zero suction do not support Bell's predictions which appear to underestimate the drag coefficient. Alternative theoretical treatments are offered by Wilson [8.27] and ESDU [8.53].

By-pass lanes are a familiar and important feature of many current designs but additional studies on by-passing are needed to produce workable correlations hopefully based on data from two-phase flow (or tubes with suction) experiments.

8.1.5 Computer Modelling

The increase of the speed of modern computers in recent years has led to a vast increase in the use of programs for solving some of the problems associated with large power plant steam condensers. In addition the advent of the desk-top computer has seen the transfer of many of the simpler mainframe programs to these portable machines; this has resulted in considerable advantages in field work where there has been no access to a mainframe.

There are two key parameters which the designer or performance predictor of power condensers requires with some accuracy and confidence, these are :
(a) the overall coefficient of heat transfer, α_0, and
(b) the overall pressure drop between entry and exit on the shell side and how this is distributed.
As indicated in section 8.1.3 it is straightforward to calculate an overall HTC where the flow behaviour is relatively obvious or is known a priori. In general, this can only be done for the case in which the flow is or can be regarded as one dimensional, or at least is well defined throughout the tube bundle. Where the design of the condenser is less than adequate and its performance is affected for example by isolated regions of the inert gases, a calculation of heat transfer can become much more complex, requiring two or three dimensional flow and heat transfer models since heat transfer becomes much more dependent on the local flow conditions. The calculation of overall pressure drop is generally quite sensitive to flow distribution and can be calculated with some degree of accuracy for only the relatively simple cases of one dimensional flow or reasonable well defined flow which can be expressed in one dimensional terms.

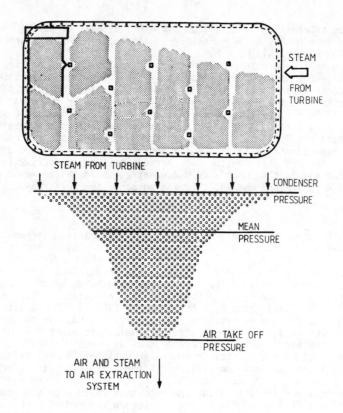

FIGURE 8.1.16 The replacement of complex nest shapes by
a one dimensional nest description function

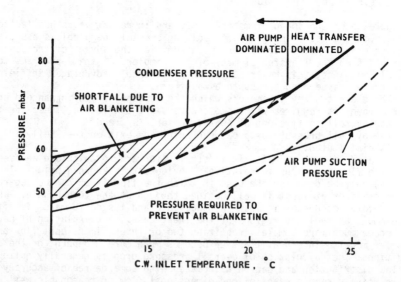

FIGURE 8.1.17 A typical 'CONE' prediction of condenser pressure [8.65]

Some condenser designs lend themselves to a one dimensional theoretical treatment and so one dimensional models are revised in the next section. The analyses of more complex tube nest designs are reviewed in section 8.1.5.2 onwards.

8.1.5.1   *One Dimensional Models*

In one dimensional models it is usually assumed that there is no variation of flow variables perpendicular to the main steam flow direction on planes perpendicular to the tubes [8.64]. Alternatively complex shapes of nest can be replaced by an equivalent one dimensional nest description function ([8.65], Fig. 8.1.16). Radial inflow geometries can also be treated as one dimensional [8.66] by examining angular segments bounded on the outside by the steam distribution lane and on the inside by a vent. For a given segment, once the steam enters from outside, it is restricted from flowing across the side boundaries to another zone. Overall performances are obtained by integrating over all zones. By dividing the longitudinal length of the condenser into a number of short bays Barsness also applied the analysis to the various bays and integrated the performance over all bays from cooling water inlet to cooling water outlet. This approach is expanded in section 8.1.5.2. A similar type of radial flow code called ORCON1 has been developed at Oakridge National Laboratory [8.63]. In these models, overall HTC was taken to be constant along the tube length.

There are fundamental difficulties to these approaches. In the Chisholm model [8.84] steam condensation, velocity, non-condensible gas concentration and inundation can be followed with a row by row calculation down through the tube nest. An arbitrary steam flow is first assumed and the correct value found by iteration until the required conditions are obtained at the vent point. It is found that with a rectangular tube bank, the influence of the walls and the introduction of by-pass lanes renders the flow strongly two dimensional on the transverse plane. For the radial inflow design, the asymmetry of performance, caused by condensate drainage, makes the calculation of pressure drop by the Barsness method difficult to handle. Overall heat transfer characteristics are, however, reasonably calculable and codes of the ORCON 1 type are used successfully.

The program CONE described in [8.65] is a one dimensional computational tool with emphasis given to the role of the air pump in condenser calculations. It will be recalled that the BEAMA/HEI codes fail to take into account tube nest layout and the interaction of the condenser with the air pump or ejector. The CONE program is a code which has been developed to overcome these particular deficiences. As shown in figure 8.1.17, the condenser pressure will be determined by heat transfer considerations alone if the air suction pressure is sufficiently low. If the air removal device does not produce sufficient suction, air blanketing will occur in the condenser and the condenser pressure will be controlled by the air suction pressure. This approach is valid for condensers where the regions of air blanketing can be reduced or removed by improved suction at the vent points. Isolated air pockets, i.e., those not connected to the vent points, are unlikely to be directly affected by improved suction and the CONE approach cannot be applied.

If the effects of isolated inert blanketing are expected to be important, one dimensional models are not suitable for predictions. Furthermore, with tube plate layouts more complicated than the simple shapes in figure 8.1.16 it becomes increasingly difficult to accurately predict pressure drop and overall heat transfer performance and higher dimension models are needed.

8.1.5.2   *Multi-Dimensional Nature of Condensers*

Air vent trunking in condensers usually runs the length of the condenser, figure 8.1.18. The pressure loss through this ducting can be a significant part of the overall condenser pressure drop. The pressure difference available to drive the steam through the condensing surface is therefore variable. This feature, together with the fact that cooling temperature increases from inlet to outlet,

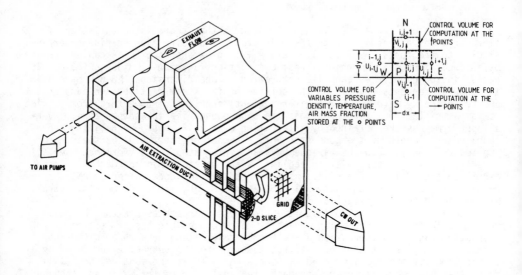

FIGURE 8.1.18 Grid structure in computer codes [8.68]

thereby causing condensation rate variation along the tubes, would lead one to
expect that the steam is highly three dimensional. In practice tube support
plates divide the condenser shell into a number of bays within which the flow can
be treated as essentially two dimensional.
     A suitable two dimensional model needs to be applied to each bay given
suitable air-ejector characteristics, a CW inlet temperature distribution and
mass flow, and a total steam flow rate (or heat load). From this information a
condenser inlet pressure (assumed uniform), an air pump pressure and an overall
heat transfer coefficient can be calculated. Two dimensional modelling techniques are described below but the whole scheme is quasi three dimensional since
starting at the bay nearest the CW inlet end, where tube side conditions are
known, calculations are performed for each bay 'marching' down the condenser to
the CW outlet end. Local cooling water temperature rises are computed and an
axial variation for the heat load is predicted.

8.1.5.3 Two Dimensional Modelling Approaches

Two approaches to modelling are to be found in the literature : nodal equations
which call for the specification of the local equations for each tube within the
bundle [8.27] and the more popular continuum model where the steam is considered
to be flowing in an isotropic porous medium which allows for the non-uniform
blockage effects of the tubes and the removal of steam due to condensation. In
view of the current understanding of the complex nature of shell side conditions,
the specification of such fine detail for the former method is perhaps unwarranted
and for large condensers can result in impracticably large computer run times.
Only the continuum modelling approach is considered here.
     Two distinct but inter-related regions need to be treated : The untubed
areas where viscous effects are small and the tube bundle areas where the steam

flow is primarily influenced by the resistance of the tubes, the condensation of the steam and the resulting condensate flow. Two fundamentally different methods expressing this problem in equation form are plausible, either to describe the total flow with a single system of equations or by separable systems of equations for the distinct flow regimes suitably matched by iterative techniques at numerical grid interfaces corresponding with the tube nest boundary. For brevity, an approach involving a single system of equations is considered in section 8.1.5.4.

8.1.5.4 Field Equations

Harlow & Amsden [8.69] and Spalding [8.70] have described successful applications of numerical schemes for solving equations of motion for two interpenetrating fluid continua. Elaborate models of multi-dimensional two-phase flow processes have since been developed generally deriving from the methods of the above authors.

The fundamental concept for two-phase prediction is that of space-sharing interspersed continua [8.70], according to which the distinct phases, steam and condensate are present within the same space, their shares of space being measured by their volume fractions r. Eleven equations governing the distribution through space are strictly required for the eleven independent variables, r, the steam velocities, u, v, w, the condensate velocities, U, V, W and the enthalpies and pressure h, H, p and P. These are provided by the conservation laws of mass, momentum and energy which must be obeyed by each phase individually, added to which is an equation governing the pressure in one of the phases as the void fraction approaches a physically obtainable limit. In addition, the idea of a phase in condensers is complicated since a distinction needs to be made between dense condensate and water droplets. To model the convection and diffusion of incondensibles, further conservation equations are needed.

Terms in the various conservation equations contain physical properties, interphase flux couplings, turbulent diffusivities etc., for which auxiliary relations are also required. It is the specification of these auxiliary equations which is important for physical realism.

Available computers can handle efficiently large sets of conservation equations and in references [8.71,8.72] there are some preliminary results from a large computer code which handles the conservation equations for steam and condensate separately applied to a simple condenser geometry. However, it is the imprecise nature of some of the auxiliary relations for condenser application, for example, turbulent diffusivity, which has led some authors to use simpler sets of conservation equations.

Much useful information and insight into condenser characteristics has and still can be gained by considering the conservation equations of only one of the phases, steam. Correction terms, based on flow regime map ideas, can usefully be added to the equations to account for two-phase flow effects. Correction factors include the slip velocity between condensate and steam in the momentum equation and the void fraction. It is this latter mathematically simpler approach which is described here. The mathematical techniques solving the more complex fully two-phase flow equations are basically no different from those to be discussed herein only the iterative strategy or sequence of operations becomes more complicated.

The two dimensional performance calculations described by Emerson [8.73], Wilson [8.77] and Ozeki [8.74] are based on a homogeneous model where steam and air velocities are assumed equal. These methods can yield acceptable results in the absence of isolated air pockets. To estimate the performance with inert gas blanketing then either the assumption of equal velocities needs to be relaxed, and the difference in velocities determined by a macroscopic turbulent diffusion law, or a similar approach to that used in the CONE program (section 8.1.5) may be applied. The errors associated with the latter approach are believed to be no greater than the former [8.65]. For completeness, equations including diffusion are presented below.

All the dependent variables of interest obey a generalized conservation principle. If the dependent variable is denoted by ψ, the general single-phase two component differential equation in a rectangular coordinate system is

$$\frac{\partial}{\partial x}(\rho u\psi) + \frac{\partial}{\partial y}(\rho v\psi) - \frac{\partial}{\partial x}\left[\Gamma_\psi \frac{\partial \psi}{\partial x}\right] - \frac{\partial}{\partial y}\left[\Gamma_\psi \frac{\partial \psi}{\partial y}\right] = S_\psi \qquad (8.1.34)$$

The variables solved for are vapour velocities in the x and y direction, u,v; the air mass fraction ϕ and pressure p. The distributed resistance method, [8.75], is used in which the velocities, u and v are those arising due to the reduced flow area in the tube nest. The variable, ψ, thus stands for u, v, ϕ or 1 where the latter corresponds to the vapour continuity equation. ρ is the vapour density. The transport coefficient Γ_ψ and the source or sink term, S_ψ are specific to a particular meaning of ψ :

Variable	Mechanism for Γ_ψ	Source Term, S_ψ
u	Turbulent viscosity	$-\partial p/\partial x + F_x$
v	Turbulent viscosity	$-\partial p/\partial y + F_y$
ϕ	Turbulent diffusion	M_e
Continuity	-	$\dot{} Q$

TABLE 8.1.2 Terms in equation (8.1.34)

In Table 8.1.2, Q is the local point value of flow condensation rate per unit volume. M_e is the mass transfer rate per unit volume of the air/steam mixture which is removed by the pump or ejector. (For condensate films which remain intact a further influential sink term for the ϕ equation might be that which accounts for an air-rich boundary layer attached to the flowing condensate [8.76]). F_x and F_y are components of a single force acting to oppose local fluid motion. If it is postulated for the present that when condensation occurs there is an immediate net loss of momentum, it can be shown that

$$F_x = -K'\rho uq - uQ \qquad (8.1.35)$$

$$F_y = -K'\rho vq - vQ$$

where q denotes the vector mean steam velocity and $|q| = \sqrt{u^2+v^2}$. K' represents a single-phase total pressure loss coefficient

$$K' = \frac{\partial P_t/\partial S}{\rho q^2} \qquad (8.1.36)$$

where $\partial P_t/\partial S$ is the pathwise total pressure gradient. (There is correspondence between K' and ξ defined by equation (8.1.33).)

The use of $\nabla \cdot (\Gamma \nabla \psi)$ does not limit the general ψ equation (8.1.34) to gradient-diffusion processes since other terms can always be expressed as part of the source term. Γ_ψ can of course be set to zero if required, viscous terms generated by local velocity gradients near tubes are still included via the force term, F. Gross viscous terms represented by the Γ_ψ terms, generate forces due to velocity gradients across the bundle as a whole and so their omission would mean solutions to equation (8.1.34) would not give rise to gross recirculation patterns. Such recirculation is not common but does occur in condenser shell side flows.

The merits or otherwise of including the turbulent diffusion terms in the equation for ϕ have been discussed earlier. Al-Sanea [8.71] provides approximate relationships for Γ_ϕ. Since data for the transport coefficients are scarce there is little experimental support for the correlations in [8.71] and the methodology needs further consideration.

To account in some way for the fact that the water flow comprises two phases, but avoiding the added complexity of computing the conservation equations for the condensate, approximate correction factors to the source terms in equation (8.1.34) are now considered :

(a) It can be assumed that the vapour which condenses gives up all its momentum to that remaining. Some authors then drop the terms, uQ, vQ in the equations (8.1.35) thereby reducing the effective frictional drag term. Frictional drag is reduced by condensation and the condensing vapour probably transfers its momentum to the vapour/air boundary layer and the water film on the tube thereby reducing the drag on the vapour/air mixture in the free stream. Alternatively,
(b) Two-phase flows can be accounted for by the inclusion of a gravitational term in equation (8.1.35b) a form of which was discussed in detail in section 8.1.4, and by substituting $\phi^2 K'$ for K' where ϕ^2 is a two phase multiplier. This gravitational term also includes a buoyancy force (since ρ_v in equation (8.1.31) can be replaced by the air-steam mixture density, ρ). The buoyancy force is included not because it would by itself be capable of affecting the steam pressure, but because its inclusion may illustrate a tendency from air to accumulate nearer the bottom of a condenser. Equation (8.1.35a) is unaffected by these modifications - for horizontal flow the gravitational term is naturally absent but also from section 8.1.4, ϕ^2 is approximately unity.

The variable, ψ, in equation (8.1.34) may also represent the stagnation enthalpy, turbulence energy, etc. and it may have been noted that the heat transport conservation equation has not been mentioned. To assimilate the large volume of data emerging from research programs on heat transfer, the popular resistance method for determining the overall heat transfer coefficient together with a conveniently defined LMTD is selected in preference to the inclusion of a further conservation equation for steam temperature. The condensation rate per unit volume, Q, to a tube or group of tubes is then computed from

$$Q = \alpha_0 \Delta T_{LMTD} \, / \, L. \tag{8.1.37}$$

8.1.5.5 *Algebraic Representation of the Field Equation*

There are many ways in which sets of algebraic equations can now be constructed to simulate the behaviour of partial differential equations. A method formalized by Spalding [8.70] has been successful for a variety of fluid mechanics problems. In this method the partial differential equations in section 8.1.5.4 are expressed in terms of finite domain algebraic equations. The algebraic equations connect variables prevailing at a cluster of nodal points arranged on a grid as shown in figure 8.1.18.

Equations are obtained by integrating the relevant differential equation over a domain of prescribed volume surrounding a central point and evaluating the resulting volume and surface integrals with the aid of interpolation functions which are presumed to govern the distribution of the variables between the nodal points. The most attractive feature of the control volume formulation is that the solution exhibits exact integral balances. Integral conservation of quantities such as mass and momentum is exactly satisfied irrespective of the number of grid points. Grid point values of ψ then constitute the solution.

There is no simple recipe for the selection of interpolation functions to evaluate the volume and surface integrals. It can be demonstrated that a simple piecewise linear profile (equivalent to central differencing) for all the ψ terms leads to unrealistic results and it may be recalled that this is why early attempts to solve convection problems by central differencing schemes were limited

to low Reynolds numbers. Most profiles are deduced from analytic solutions of
reduced forms of the governing equations. One useful form is upwinding of the
convective terms combined with a power law adjustment of the diffusion term to
give a better behaviour at large Peclet numbers.

Since the equations (8.1.34) exhibit both nonlinearity and coupling, iterative procedures must be employed. Much experience with single-phase flows has
led to the development of many procedures. Influential codes based on the early
work of Chorin [8.77] are MAC [8.78] and SIMPLE [8.79] with its variants SIMPLER
and SIMPLEST. The various methods differ according to the handling of the coupling between the equations, the number of variables to be updated simultaneously,
and in what order. Details of these various methods are not given here as they
are well documented elsewhere. Also the resulting set of linearized algebraic
equations can be solved by a variety of means. These are not discussed either,
except to note that Stones SIP procedure has proved to be computationally efficient for condenser calculations [8.68].

Finally, Shida et al. [8.80] point out that the techniques that have been
described are restricted to condensers with simple shapes because of the assumed
rectangular mesh pattern and finite domain technique. This is not entirely
accurate. The techniques described do not preclude the use of non-uniform meshes
and, as will be seen in section 8.1.6, the condensers which have been analyzed
are by no means simple. Nevertheless, the mesh of Shida gives potentially better
definition of nest boundaries, drainage trays, baffles, etc. than rectangular
grids. The mesh is based on a triangular pattern conventionally used in the
finite element method. The solution technique is the Los Alamos fluid in-cell
method and they perform calculations using upwind two-step Lagrangian and Eulerian
time-marching techniques. This is an interesting development area as is the application of the finite element technique to condenser calculations [8.74].

8.1.6   Examples of Computer Studies

8.1.6.1   Computer Program Validation

Experience on the accuracy of the one dimensional CONE program is being gained by
comparison with tests on plant. An example is given in figure 8.1.19 showing

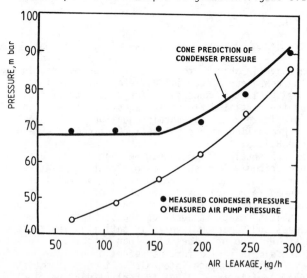

FIGURE 8.1.19 Measurements and 'CONE' predictions for the effect
of air leakage on condenser pressure for a 500 MW condenser

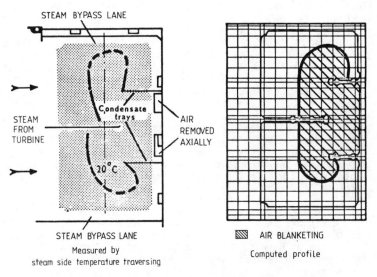

FIGURE 8.1.20a Measured and computed regions of air
blanketing in a section of a 500 MW condenser

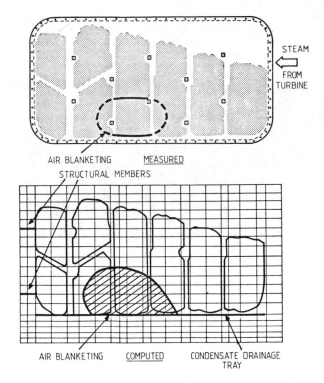

FIGURE 8.1.20b Measured and computed isolated
air pocket in part of a 500 MW condenser

the effect of air leakage on condenser pressure for a 500 MW condenser.
      As shown, the condenser pressure is initially constant indicating that pro-
viding air does not stagnate, it does not affect heat transfer coefficients
within a tube nest. When the air leakage rate is high enough, the air pump pressure becomes too high for complete ventilation. Air pockets form and the condenser
pressure rises to match the air pump pressure. This same effect was observed
irrespective of where the air was injected. This evidence tends to support the
modelling approach used in CONE in which the degree of air blanketing in a condenser is defined by the vent pressure. The role of air leakage is to solely
determine the vent pressure.
      As the basic heat transfer calculation within the CONE program is based on
an empirical relationship similar to that defined by the BEAMA and HEI recommendations, the accuracy of the program for general use should be comparable with
these documents if the steamside pressure loss is small and better when tubes are
more tightly packed within the condenser shell and steamside pressure losses are
not negligible.
      The CEGB quasi three dimensional program EPOC [8.68] has been applied to a
number of condenser geometries. Two particular condensers, described in further
detail in [8.65] are shown in figures 8.1.20a,b. These figures demonstrate the cap-
bility of the program to predict regions of air blanketing, whether isolated or
not. The agreement between predicted and measured performance, although not perfect, was considered satisfactory for the purpose of evaluating modifications to
these condensers.
      Because of the uncertainties regarding the mechanism by which air accumulates
a simple air prediction model was used for these studies. Extending the CONE
principle, no air conservation equation was satisfied and it was assumed that a
sharp boundary exists between regions of high and low air mass fractions with no
condensation taking place in the blanketed regions. This is supported by steamside measurements on condensers where a rapid transition has been observed. The
location and size of air pockets is determined instead by an isobaric boundary
condition at the air/steam interface.
      Experimental evidence to support pressure drop and heat transfer coefficient
predictions is rather limited. In [8.65] the heat transfer correlations in
figure 8.1.12 were used in a model to compare predictions of local heat transfer
coefficients with those deduced from measurements of a condenser shown in figure
8.1.21. Computed results show good agreement in the top half of the nest but
underestimate in the bottom and front of the nest. Explanations are offered in
[8.65] for the differences and the application of more recent correlations may
help to resolve the disparities. No measurements were made of pressure variation
but computed values of pressure drop through the nest agree quite well with
values inferred from steam temperature measurements.

8.1.6.2   Modelling of Recent Designs

An example of an application of EPOC is shown in figure 8.1.22. This shows the
distribution of heat transfer coefficient computed for the tube bundles in a
large transverse underslung condenser recently analyzed. A comparatively even
condensation rate is predicted across the tube bundle with no indication of isolated air pockets. Exact performance details are not available but preliminary
information suggests that the high values of overall heat transfer coefficients
are being obtained.
      The program EPOC was used at the design stage to assess the anticipated per-
formance of the UK-Ince 'B' pannier condensers mentioned above and shown in
figure 8.1.21. Recent measurements indicate that these condensers, with four
individually vented tube nests, are producing a vacuum which is 4 mbar better
than guarantee with a terminal temperature difference of 3.4°C.
      Similar codes are used by other utilities for design purposes. The program
CALICO has been developed by Electricité de France [8.81]. The detailed flow
patterns have proved useful to optimize geometrical parameters with respect to

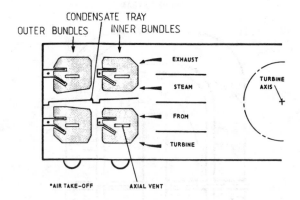

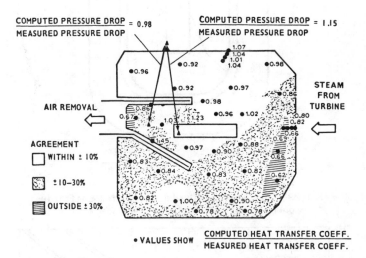

FIGURE 8.1.21   Comparison of computed and measured values of
heat transfer coefficient and pressure drop

vibration problems. Toshiba Corporation [8.74] has a design method based on
the finite element method. The method assumes vortex-free potential flow in untubed regions of the condenser and the agreement between predictions and hydraulic analogue experiments for current bell-type condensers seem to support the
potential flow assumption.

8.1.6.3   *Improvements to Existing Condensers*

Tube cleaning and attention to air inleakage and extraction pump performance
are routine maintenance tasks which can increase operating costs if not timed
optimally. The method of on-load measurement of tube cleanliness is a tool which
can give the current state of the plant. Program CONE can predict the performance
gain from tube cleaning and can also indicate the tolerance levels of air inleakage before performance falls off. Figure 8.1.23 shows a characteristic calculated by CONE for a 500 MW condenser which shows for example that less air leakage
can be tolerated at lower cooling water inlet temperatures and when the tubes are
clean. CONE predictions can also be compared with operating data to check for
mal-operation of plant.

EXHAUST STEAM

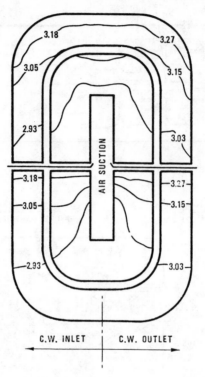

FIGURE 8.1.22 Computed overall heat transfer coefficients
on CW inlet and outlet planes (KW/m²K)

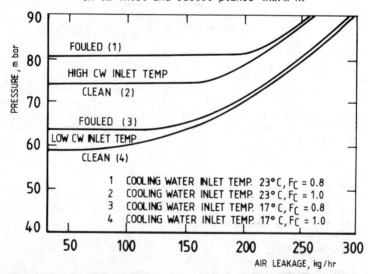

FIGURE 8.1.23 Prediction of the effect of air leakage on condenser
pressure for different cleanliness factors and CW inlet temperatures

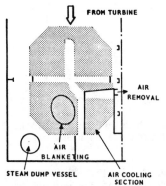
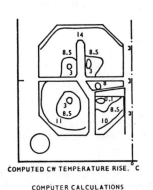
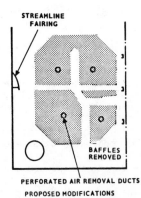

FIGURE 8.1.24 An example of the use of computer codes to assess the effect of modifications to existing plant to improve performance

Colquhoun-Lee & Bryan [8.82] have computed flow patterns in underslung condensers in the UK. The predictions indicate routes of unequal resistance to the air removal points. Suitable baffles were added to prevent steam taking the line of least resistance and swamping the air ejectors.

Modifications to the Magnox station condensers at Oldbury (Fig. 8.1.24) are described in some detail in [8.3, 8.65]. Briefly, the problem was not steam by-passing but very deep tube nests with a high steamwise pressure loss. The task was to reduce the average path length. Computed cooling water temperature rises in figure 8.1.24b show regions of poor condensation. Venting at the centre of the four tube nests was the obvious and most beneficial answer but the real use of the computer program was to help decide whether venting the top nests as well as the bottom gave sufficient improvement to warrent what were difficult modifications to implement. Computer checks were also made to see if extra access lanes would be beneficial and whether condensate trays should be provided below the upper nests. Engineering constraints in any event proved to be the limiting factor and modifica tions for the lower nests were only made. Two central suction ducts were created and the baffles removed as shown in figure 8.1.24c. Preliminary tests carried out after the modifications have indicated an improvement in performance.

8.1.7 Final Remarks

This review is by no means exhaustive, notable omissions are the studies on the performance characteristics of enhanced condenser tubing (roped tubes) [8.3,8.83, 8.84]. This is a most obvious candidate for giving immediate performance improvement and the CEGB considering its potential in line with other utilities.

Condenser plant improvement is a continuing process and considerable progress has been made in the wake of the oil crisis in the early 1970's. Progress will continue to be made since there are many gaps left in our knowledge of the condensing plant behaviour and so more efficient conversion of energy is still possible. With present computer programs the accuracy of pressure drop predictions is believed to be about 2 mb in absolute terms. Relatively between computations for different data, or small geometrical changes, the accuracy should be better, probably of the order 1 mb.

ACKNOWLEDGEMENT

The author thanks many colleagues within the CEGB and the manufacturing companies whose work forms the backbone of this review.

8.2 STEAM CONDENSER DEVELOPMENTS

H.V. Lang

8.2.1 Introduction

Steam condensers enable the operation of steam power plants in closed water/steam cycles. Such closed cycles are characterized by expansion of steam in low pressure turbines to a pressure below ambient pressure and by conservation of the working fluid.

The power plant condenser acts as a heat sink. There are basically two types of condensers with respect to where condensation of steam takes place :
- surface type condenser,
- direct contact condenser.

Both condenser types can be cooled by water or directly by air. With respect to the pressure at which condensation takes place condensers can furthermore be separated into :
- vacuum condensers,
- atmospheric condensers,
- heater/condensers for district heating.

Because of their importance and dominance in utilization, this paper is limited to water cooled, surface type vacuum condensers. For the preparation of this text extensive use was made of the condenser technology of Brown Boveri & Company, Ltd., accumulated by development, design, commissioning and maintenance of power plant condensers of all sizes for more than 80 years. Where appropriate and accessible reference will be made to technologies of other condenser suppliers.

8.2.2 Condenser Development Goals

The ideal power plant condenser should exhibit the following characteristics :
- it condenses steam at a temperature that closely approaches the cooling water exit temperature;
- it discharges condensate at a temperature that is identical to the condensation temperature;
- it discharges condensate free of impurities like oxygen, carbon dioxyde, or substances leaking into the condenser from the cooling water;
- it subjects the cooling water to minimal pressure drop;
- it acts as a perfect dump for low and high energy fluids;
- it requires minimal vent pump capacities;
- it operates during the life time of the power plant without being responsible for forced outage of the plant or increased unavailability of the plant;
- it operates in the above sense with minimal effort for inspection and maintenance;
- it fits into the space under the low pressure turbine with minimal effect on turbine foundation and foundation height;
- its first cost and the cost for erection and commissioning are minimal.

Contrary to superficial judgement the condenser is not a simple shell and tube heat exchanger that can be assembled by every shop if it is to meet the above characteristics.

The Electric Power Research Institute (EPRI), the Empire State Electricity Energy Research Corporation (ESEERCO) and the Edison Electric Institute (EEI) have spent US $ 10 M within the last five years for research in the area of improvements in US steam surface condensers. A survey performed by these three agencies in the USA showed that for fossil-fueled plants of 600 MW and greater, the loss of unit availability directly attributable to condensers is 3.8% [8.91]. In nuclear-fueled plants, the direct loss attributable to condensers is probably of the same magnitude. Problems related to condenser reliability cost the electric utility industry $ 600 M annually for replacement power alone.

European condensers by and large have good performance because, in the majority of plants, condenser and turbine were supplied together, tying the condenser performance into the turbo set heat rate and bid-evaluation at the tendering stage. This and the somewhat greater energy consciousness in the past (government controlled price of electricity) promoted the development of improved condenser concepts.

8.2.3 Factors Determining the Condenser Size and Tube Arrangement

The principal factor that determines the size of the condensers, i.e., the surface area required to condensate a given steam flow rate at a specified temperature, is the heat transfer rate. This rate, however, has to reflect the thermal performance of a plurality of tubes, each, as a member of a tube pattern, having its individual boundary conditions. These individual conditions govern heat transfer as well as other performance characteristics like deaeration, condensate depression and venting requirements. It is therefore necessary to identify and understand the physical processes in a condenser. Since the basic equations governing heat transfer and steam side pressure drop are dealt with in chapter 8.1 only the effects and consequences of some of the major processes are covered here. The selection is made on the impact these processes have on the bundle shape and condenser size.

8.2.3.1 *Steam Side Pressure Drop*

Steam penetrating the condenser bundle is subject to pressure drop because tubes, tube arrangement and condensate act as flow resistances. The local condensation temperature thereby will be reduced, decreasing the temperature difference to which the amount of heat transferred is proportional. Consequently the overall heat transfer rate will drop.

This drop was observed while testing large condensers whose design had been extrapolated from those of the 1950-60's. At that time Mr Dunham of TVA observed [8.92] that "modern large condensers should not be expected to have performance characteristics as indicated by HEI [8.4] recommended heat transfer rates". He concluded that this was because of steam side pressure drop since the heat transfer rates approached or even surpassed the HEI values while throttling the cooling water flow. Throttling the cooling water flow results in higher condenser pressure at which the effect of pressure drop is less pronounced. To illustrate this effect : at a condenser temperature of 30°C a pressure drop of 3 mbar corresponds to a temperature drop of 1.3 K whereas at a condensation temperature of 60°C the same pressure drop corresponds to a temperature drop of just 0.3 K.

Numerical simulation at Brown Boveri [8.94] of a tube bundle section of constant inlet cross section but varying number of tube rows (Fig. 8.2.1) and variable tube arrangement revealed that not only is the absolute amount of pressure drop important but also the location where most of the pressure drop occurs. The simulation indicated that the absolute rate of condensation decreased for this tube bundle when the steam supply to it was increased for fixed steam inlet temperatures, regardless of the compensation by tubes added to the rear of the bundle. The reason is that the pressure drop increased in the first tube rows to such an extent that the driving temperature difference disappeared for the rest of the tube rows. It should be noted that the condenser pressure may be governed by the pressure at which the vent pump draws off the mixture of steam and incondensible gases. The incondensibles to the largest part consist of air inleakage into all vessels whose pressure is below ambient, the rest being made up of gases decomposed from chemicals used in water treatment or of gases dissolved in the makeup water. The condenser pressure thus equals the vent pump pressure increased by the steam pressure drop in the air cooler and in the condenser proper. A bundle with low pressure drop will therefore enable lower condenser pressures for a fixed vent pump

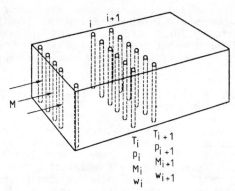

FIGURE 8.2.1   Tube bundle section for numerical simulation of steam
flow and steam condensation

capacity and will actually reduce the steam loading of the vent pump. This correlation of condenser pressure and vent pump capacity is important but is usually
overlooked. Figure 8.2.2 shows the result of a theoretical analysis performed by
Brown Boveri for their 'church-window' condenser. These results have been vali-
dated by field tests. Reduction of turbine load will lead to lower condenser pres-

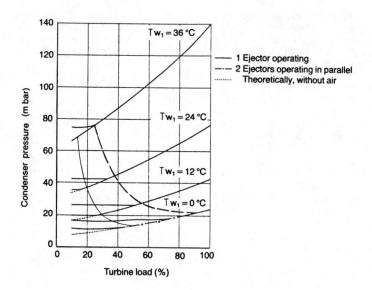

FIGURE 8.2.2   Effect of vent pump capacity on condenser
pressure at various loads

sure as long as all incondensibles can be removed from the condenser by the vent
pump. Otherwise incondensibles are backed up into the bundle, reducing the effective heat transfer area. Then the condenser pressure is governed by the vent
pump suction pressure, as is the case in figure 8.2.2 below 90% load with the
given vent pump and cooling rates temperatures of 0°C. Doubling the vent pump
capacity lowers the vent pump governed load range to below 50% load.

We have considered the pressure drop as steam penetrates the bundle. Pressure drop ahead of the bundle between turbine exhaust and bundle periphery has the same detrimental effect of lowering the condensation temperature and has to be minimized in order to best utilize the temperature level of the cooling source.

In summary high heat transfer rates can be approached only with bundle concepts that lead to small pressure drop. However, since space required by the condenser will effect the cost for the machine house building, the heat transfer rates, pressure drop and bundle concept are the result of an optimization process. Numerical investigations of different tube arrangements [8.93] for the same condenser load showed that there is an optimum number of tube rows to be penetrated, that is slightly dependent on cooling water inlet temperature.

In order to develop such optimal tube bundle concepts Brown Boveri [8.93, 8.94] devised a laboratory test method to simulate heat transfer and pressure drop (Fig. 8.2.3). In the test the working fluid steam is replaced by water, condensation on tubes by water draining into perforated tubes, heat transfer by flow resistance of this drained water in capillary tubes (length and diameter of the capillary tube adjusted to simulate the heat transfer resistance of the tubes) and temperature difference by level difference in inlet and level tanks. The levels in the latter were adjusted for the difference in "heat transfer" - equivalent

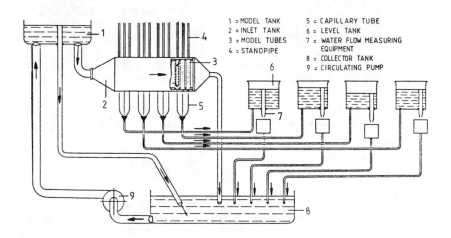

FIGURE 8.2.3 Diagrammatic arrangement for tests on a model condenser with simulated flow and heat transfer

to the water drain rate from a group of tubes - while successively simulating slices of the condenser from cooling water entry side to cooling water exit side. This laboratory test method was calibrated using field test data. The first generation Brown Boveri 'church window' concept (Fig. 8.2.4) was developed by this laboratory method for optimum pressure drop, and with the optimum number of tube rows.

8.2.3.2 Incondensible Blanketing

Incondensible blanketing - usually referred to as air blanketing - occurs whenever the steam flow rate through a region of the bundle is equal or less than the rate of condensation within this region. As a result incondensibles transported with the steam are deposited at these zones and the effective heat transfer area is reduced accordingly. Without the presence of incondensibles such regions would show up as pressure sinks in the mapping of absolute pressures for a cross section of a bundle. This was first observed while testing the first generation 'church window' bundle (Fig. 8.2.5) in the test rig (Fig. 8.2.3).

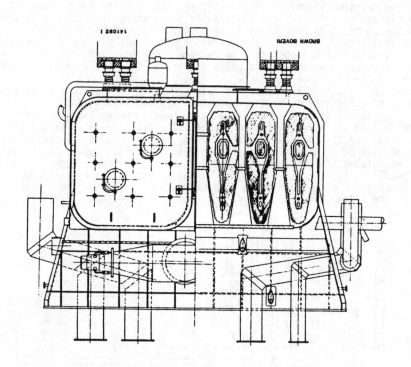

FIGURE 8.2.4 First generation 'church window' condenser for
Asnaes III, Denmark, commissioned 1968

The mapping of pressures clearly shows such low pressure zones, i.e., closed
isobars containing lower pressures, to the left and below the air cooler of the
right-hand bundle. The significance of this became evident, however, when these
same areas were seen to have the smallest cooling water temperature rise in the
mapping of these temperature rises (Fig. 8.2.6) obtained in the field.

The same test rig also indicated that steam supply lanes left untubed in bundles to increase the bundle periphery and thereby reduce pressure drop, can actually promote the formation of such pressure sinks (Fig. 8.2.7).

As already mentioned such low pressure zones trap incondensibles. Figure 8.2.8
depicts why then such a zone is rendered ineffective for heat transfer [8.95].
The partial pressure of steam and hence the local condensation temperature drop
according to Dalton's law when steam condenses, while the incondensible concentration increases. This incondensible blanketing lowers or eliminates the driving
temperature difference and should therefore be avoided. This requires the shaping
of the tube bundle containing an optimum number of tube rows such that there are
no peripheral steam supply lanes and that the resulting isobars are concentric
with respect to the air cooler (Fig. 8.2.9). The latter is designed specifically
to act as a pressure sink within the bundle. Then every section of the bundle is
purged adequately while the steam/incondensible mixture with slowly increasing
concentration of incondensibles converges towards the air coolers, from which it
is vented. The flow field (Fig. 8.2.10) implies that the bundle is vented into
the air coolers continuously over the full length of the air cooler.

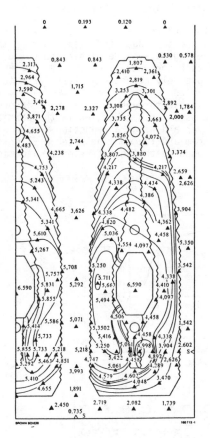

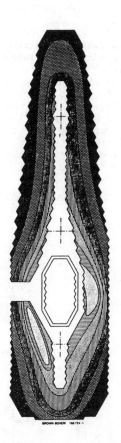

FIGURE 8.2.5 Laboratory testing of the first generation 'church window' condenser, pressure mapping

FIGURE 8.2.6 Field testing of first generation 'church window' condenser, mapping of cooling water temperature rise

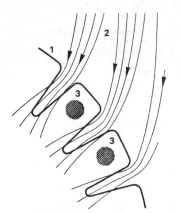

1 - Bundle periphery
2 - Steam supply lane
3 - Low pressure zone

FIGURE 8.2.7 Effect of steam supply lanes at the bundle periphery

8.2.3.3 *Effect of Incondensibles on Heat Transfer*

The condenser has to be designed to condense as much of the steam as possible to minimize makeup water requirements and to reduce the steam load on the vent pump. Thus even when incondensible blanketing is avoided within the condenser proper, incondensibles will be present in such concentrations towards the end of the condensation path that they will adversely affect heat transfer rates by depositing around the tubes while the steam condenses.

Incoming steam first has to diffuse through this layer of incondensibles. According to Dalton's law this necessitates a reduction in the partial pressure of steam towards the condensate layer and hence a reduction in the effective temperature difference between the steam and the cooling water. Essentially this boundary layer phenomenon can be compared to the macroscopic situation as depicted in figure 8.2.8.

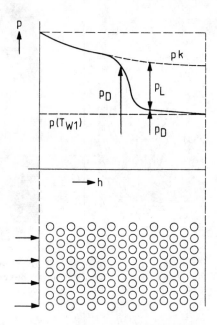

p_K condenser pressure, p_D partial pressure of steam, p_L partial pressure of incondensibles, $p_{(T_{W1})}$ saturation pressure corresponding to the cooling water inlet temperature, h depth of bundle

FIGURE 8.2.8 Development of pressures while condensing
steam/incondensible mixtures

Analysis has confirmed that a high flow Reynolds number tends to decrease the incondensible boundary layer thickness and to reduce the drop in partial pressure of the steam through the layers. The gain in heat transfer rates is pronounced for intermediate incondensible concentrations but levels out for high concentrations. Therefore Brown Boveri introduced special precoolers upstream of the air coolers in their improved 'church window' condenser concept (Figs. 8.2.9 and 8.2.10).

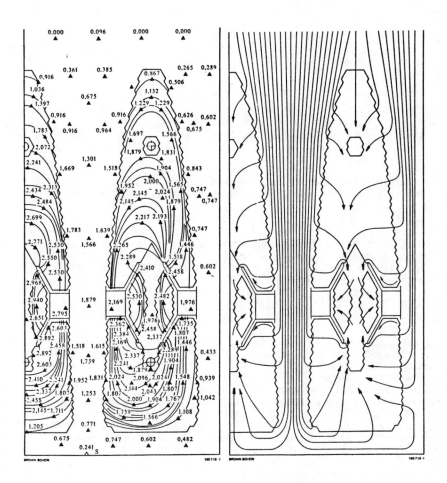

FIGURE 8.2.9 Laboratory testing of improved 'church window' bundles, pressure mapping

FIGURE 8.2.10 Laboratory testing of improved 'church window' bundles, steam flow field

8.2.3.4 Vent Requirements

Steam that has to be vented off together with incondensibles has to be compressed to ambient pressures and therefore affects the incondensible capacity of the vent pump. This steam furthermore has to be replaced by makeup water. Accordingly the condenser has to be designed to minimize the amount of steam vented off. The absolute minimum is given in accordance with Dalton's law when the partial pressure of the steam at the air cooler outlet reaches the saturation pressure corresponding to the cooling water inlet temperature, though the amount of steam additionally depends on the total pressure of the mixture. This difference in total pressure and partial pressure is commonly expressed in degrees of subcooling by forming the difference between the corresponding saturation temperatures. The air cooler has then to be sized to achieve a certain subcooling.

It should be noted, however, that the concept of air cooler design to a certain subcooling and selection of the vent pump accordingly has severe drawbacks :

- It does not reflect the actual condenser specification, i.e., cooling water temperature rise and approach of condenser temperature are not accounted for.
- It does not reflect low load operating conditions.
- It is valid for the design point and the design incondensible flow rate only.

To illustrate the last statement : reduction of the inleakage rate of incondensibles in the field will result in lower vent pump suction pressures and hence higher steam vent rates. Accordingly "subcooling" will be reduced, implying worse performance of the condenser air cooler. In reality the condenser will be vented more effectively and will exhibit better thermal performance.

At Brown Boveri the first two drawbacks of the "subcooling" concept are avoided by sizing the air cooler such that the outlet mixture temperature closely approaches the cooling water inlet temperature over the entire operating range.

8.2.3.5 Deaeration

Oxygen dissolved in the condensate enhances corrosion of metallic surfaces, and its concentration should therefore be kept as low as possible when operating the feedwater circuit with pH-values in excess of 9.

Oxygen is added to the condensate by absorption from the oxygen rich incondensible boundary layer discussed in §8.2.3.3. According to Henry's law the concentration of oxygen in the condensate is proportional to the partial pressure of the oxygen in this boundary layer. Any measure that maintains high heat transfer rates in the presence of incondensibles is therefore beneficial in keeping oxygen concentration low.

Oxygen is also added with the makeup water, since this water is saturated with oxygen due to the makeup water treatment process. This makeup water therefore has to be deaerated, a process which again benefits from all provisions for high heat transfer.

Finally oxygen can be added to the condensate by air leaking into the condensate at the condenser floor, the hotwell or the condensate pump. Adequate design -for example replacement of bolted tube sheets by tube sheets welded to the shell- and adequate maintenance should help prevent this.

It is known that some condenser suppliers doubt the deaerating performance of their condenser concepts and add deaerating hotwells. Brown Boveri has discontinued to do so for more than 10 years because they experienced extraordinary low oxygen concentrations under normal operating conditions and very inefficient low pressure thermal deaeration under abnormal conditions. Certainly the best approach is to avoid or limit oxygen build-up in the condensate from the start.

8.2.3.6 Condensate Depression

Condensate subcooling can be divided into two types. One is the subcooling of the condensate layer around the tubes, unavoidable since heat transfer is possible only when a temperature difference exists. This subcooling, however, is small. The other kind of subcooling results from the difference between the reference pressure, i.e., the pressure at the condenser inlet and the local saturation pressure. This difference originates from either a pressure drop or a high incondensible concentration. As with deaeration, the best approach to low condensate depression is optimized heat transfer performance although, should subcooling exist, condensate "reheat" is a practicable way to further reduce subcooling. Due to the fact that the pressure at the condenser floor can be higher than the condenser pressure above the tube bundle because of recuperation of dynamic pressure, condensate temperature can be even higher than the saturation temperature corresponding to the condenser pressure above the tube bundle.

8.2.3.7 Cooling Water Velocity

It can be shown that the overall heat transfer rate is proportional to the square root of the cooling water velocity. On the other hand cooling water pumping

power is proportional to the square of this velocity. Increase in cooling water
velocity tends to alter the condensers towards less width but greater tube length.
Tying these factors together results in an optimal cooling water velocity that
ranges between 2.3 and 2.6 m/s depending on equivalent cost of energy, tube material and space penalization. However, there are certain tube materials like copper alloys that are known to show inlet impingement attack at higher cooling water
velocities, limiting the velocities to about 2.2 m/s for those tube materials.

8.2.3.8 Tube Material Selection

The resistance to heat transfer through the tube wall according to Fourier's law
of conduction is proportional to wall thickness and inversely proportional to
thermal conductivity of the tube material. This resistance can be considerable.
For example the use of stainless steel instead of brass will reduce the overall
heat transfer rate by more than 20% for wall thickness of 18 BWG (= 1.245 mm).
On the other hand the tube material has to be adequate for the site cooling water
quality. Because of pollution there is a remarkable shift from the formerly pre-
dominantly used brass tubing towards more noble materials like stainless steel
and titanium.

Heat transfer characteristics, technical limitation of cooling water velocity and corrosion behaviour with respect to the cooling water, all have to be
considered for the specification of the tube material.

8.2.3.9 Tube Diameter Selection

Single tube heat transfer coefficients fall slightly with increasing diameter. With
increasing diameter, the number of tubes decreases for a given cooling water flow
rate, resulting in less cost of drilling the tube sheet and sagplates and of
joining tube sheet and tubes. Furthermore the single tube heat exchange surface
is proportional to the outside diameter whereas the cooling water flow cross
section is proportional to the square of the inside diameter. Hence the overall
dimensions of the condenser will be affected by the tube diameter. Finally, wall
thickness has to be adequate for manufacture and operation. As with cooling water
velocity and tube material the diameter selection has to be done by an optimization process. Currently the majority of tube outside diameters lie between 22
and 25.4 mm with extreme cases of 19 or 33 mm. Wall thickness varies between 1.2
mm (for copper alloys) and 0.5 mm (for titanium).

8.2.3.10 Fouling

In actual operation the heat transfer surfaces become fouled. For the cooling
water side one has to consider macrofouling and microfouling.

The term macrofouling applies when larger objects plug the tubes. These
objects can be foreign matter introduced with the cooling water in the absence
of or with defective inlet screens or they can be organisms such as mussels that
pass the screen when small and then settle to grow downstream of those screens.
Whenever these organisms die or are flushed away they can be introduced into the
waterbox and block tubes. Backwashing is one effective method of controlling
macrofouling.

Microfouling is the buildup of a fouling layer made up of tube material
corrosion products and of matter deposited on the tube wall. Microfouling can be
limited by in service cleaning methods using sponge balls introduced into the cooling water or brushes in cages at each tube inlet or outlet end that are forced
through each time a flow reversal takes place. Alternatively it can be limited
by mechanical cleaning during shutdown periods of a part or the whole condenser.
The effect of such fouling layers on heat transfer rates can be considerable as
is depicted in figure 8.2.11.

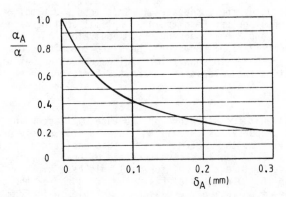

FIGURE 8.2.11 Effect of fouling layer thickness on heat transfer
λ_A = 0.2 W/mK is the assumed conductivity of the fouling layer

The steam side tube wall also becomes covered with matter that is carried
along by the steam. This is limited to tubes at the periphery and under normal
conditions does not affect heat transfer markedly.

In summary fouling does not depend on bundle concept but on tube material,
cooling water intake structure, provisions of intake screens and filters and
cleaning systems. Usually fouling is compensated during design by the addition
of 10 to 15% of excess heat transfer surface.

8.2.3.11 *Enhanced Heat Transfer Tubes*

There are a number of proposals to enhance shell or tube side heat transfer rates.
For example it was proposed [8.96] to use swagged helical or multifluted tubes
that should increase condensing steam side coefficient by six times and the tube
side coefficients by 2.5 times. Another suggestion was to make use of capillary
forces [8.97] to drain away condensate and bring down the mean condensate layer
thickness. The reasons these proposals have yet to be used significantly for
steam power plant condensers are several :
- steam side heat transfer coefficient is too large as to affect overall heat
transfer rates by its improvement;
- fouling of capillary grooves renders them ineffective;
- erosion/corrosion of tubes is likely downstream of indentations by cooling water
containing particulate impurities such as, for example, sand;
- difficulty of in service and out of service cleaning of the fouled cooling
water side.

8.2.4 Tube Nest and Condenser Concepts

In this chapter a selection of tube nest shapes had to be made on the basis of
availability to the author of drawings from publications and sales brochures.
When known to the author these propriety concepts are explained but not judged
or commented on. These concepts should be judged on their performance as determined in field tests.

8.2.4.1 *Brown Boveri & Company, Ltd. Condenser Concepts*

The 'church window' concept

In section 8.2.3 the Brown Boveri 'church window' bundle was already introduced
(Figs. 8.2.5, 8.2.9, 8.2.10). The actual 'church window' condenser (Fig. 8.2.12)

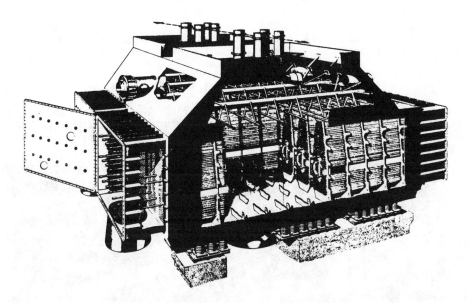

FIGURE 8.2.12   Brown Boveri 'church window' condensers

is made up of 4 to 16 of these bundle modules of identical size each.   In addition to the variable number of modules, a line of modules of different size, i.e., of different number of tubes per bundle, were developed according to the same design principles.   This allows appropriate selection of condenser size for the full range of unit output (from 100 to 2000 MWe) and operating parameters. About 170 of such 'church window' condensers are in service around the world since 1968.

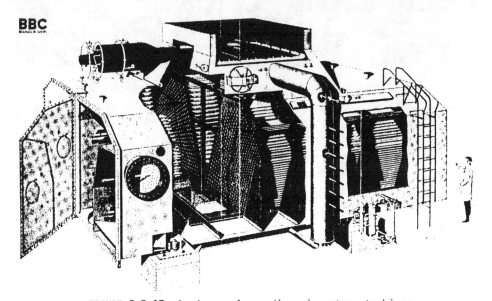

FIGURE 8.2.13   Condenser for medium size steam turbines

The CV-condenser concept

For medium size plants (below 110 MWe) Brown Boveri employs a condenser with variable pitch (Fig. 8.2.13) that consists of two identical bundles of 1 to 3 cooling water passes each. A number of bundle sizes have been developed.

The CO-condenser concept

This condenser concept (Fig. 8.2.14) is used for small plants of below and around 20 MWe and for auxiliary condensers for feedpump turbines. This concept has been used since about 1900 and in more than 330 installations since 1958.

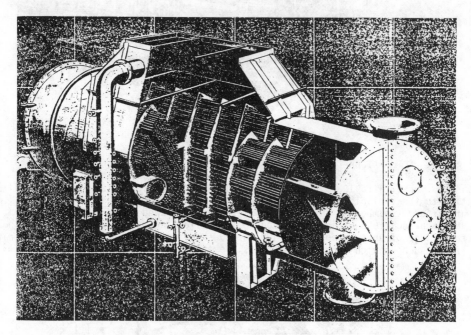

FIGURE 8.2.14 Condenser for small size steam turbines

8.2.4.2 Delas-Weir Condenser Concept

Figure 8.2.15 [8.98] shows an individual Delas-Weir tube nest. A condenser [8.99] would be made up of a number of these bundles (Fig. 8.2.16).

8.2.4.3 Ecolair Condenser Concept

This bundle [8.98], originally of Ingersoll-Rand design, is shown in figure 8.2.17 Two bundles are contained in one condenser shell.

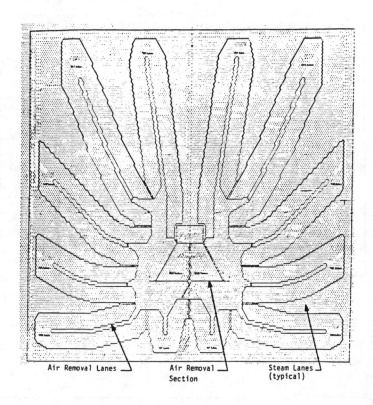

Air Removal Lanes — Air Removal — Steam Lanes —
Section (typical)

FIGURE 8.2.15 Delas-Weir tube bundle

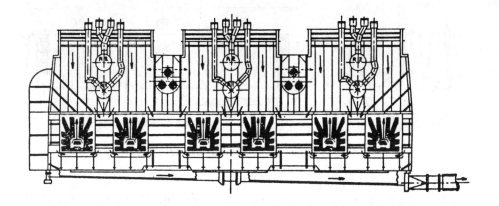

FIGURE 8.2.16 Delas-Weir condenser

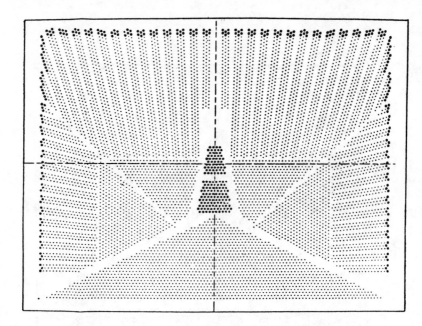

FIGURE 8.2.17 Ingersoll-Rand/Ecolair tube bundles

8.2.4.4 *Foster-Wheeler Condenser Concept*

Figure 8.2.18 depicts a Foster-Wheeler condenser design [8.100] here for a multi-pressure condenser.

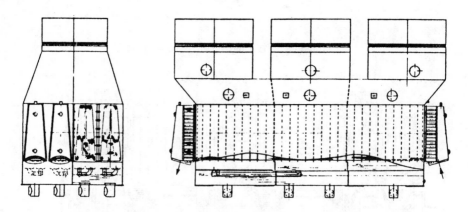

FIGURE 8.2.18 Large triple pressure single shell condenser

8.2.4.5 *KWU Condenser Concepts*

Figure 8.2.19 depicts the condenser bundle of KWU [8.101] used for fossil fired and pressurized water reactors.

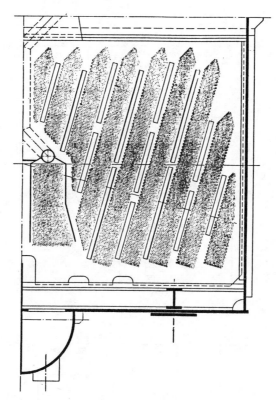

FIGURE 8.2.19   KWU-"KASTEN"-condenser

For nuclear BWR-units an alternate design is used (Fig. 8.2.20).

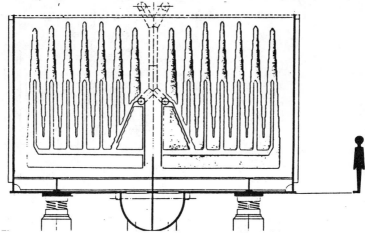

FIGURE 8.2.20 KWU-"KASTEN"-condenser for BWR installations

For power plants up to 150 MWe KWU use their "round" condenser concept (Fig. 8.2.21).

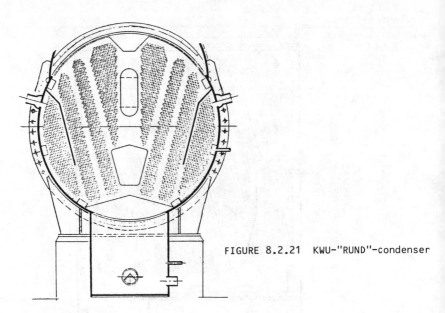

FIGURE 8.2.21   KWU-"RUND"-condenser

8.2.4.6   MAN-Condenser Concepts

For large power plants MAN [8.102] use a concept as depicted in figure 8.2.22.

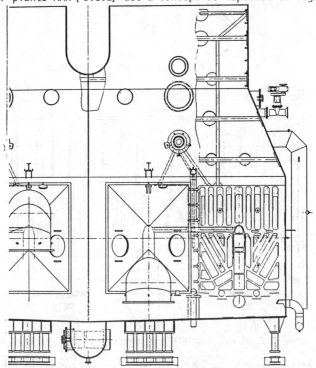

FIGURE 8.2.22   MAN-condenser for large power plants

For smaller plant size a different concept is used (Fig. 8.2.23) shown here with two cooling water passes.

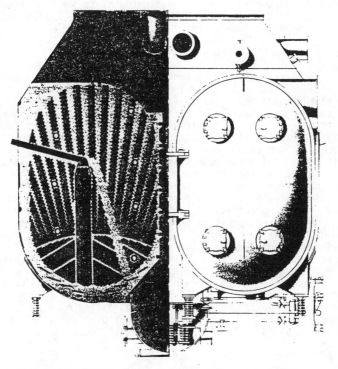

FIGURE 8.2.23 MAN-condenser for small power plants

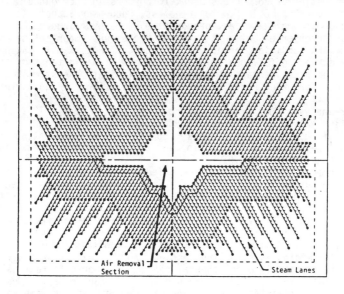

FIGURE 8.2.24 Westinghouse/Marley tube bundles

8.2.4.7 Marley-Condenser Concept

Their concept [8.98], originally of Westinghouse, is shown in figure 8.2.24. Two
to four bundles are contained within one shell. In the latter case the tube
bundles are arranged in double deck.

8.2.5 Thermal Design

A condenser is always designed to condense a certain amount of steam at a certain
condenser pressure and to transfer the heat to a cooling water of certain flow
rate and inlet temperature. The lower the pressure at which the steam is condensed
the lower the expansion end point in the turbine and hence the greater the energy
converted from the steam. Since lower condenser pressure corresponds to smaller
temperature differences with respect to cooling water temperatures, more heat
transfer area for the same amount of heat transferred is required. Again there
is no best design point but rather an optimal design point. This optimization
has to include the turbine, the condenser, the machine house, the cooling water
system and the energy produced or needed for auxiliary power.
     The optimization process would be unrealistic without the knowledge of the
overall heat transfer coefficient that corresponds to each condenser design concept and to each set of operating parameters. In chapter 8.1 the HEI and BEAMA
prediction methods were introduced. In principle these predictions base overall
heat transfer rates on cooling water velocity, cooling water inlet temperature,
tube diameter and tube material. Further parameters like condenser concept, condenser size, tube arrangement, ligament thickness, steam loading, cooling water,
temperature rise, tube length, incondensible loading and vent pump capacity do
not affect these predictions. This should be kept in mind when using these prediction methods. They should serve as a rough estimate only. The condenser
supplier should know best how his condenser performs and should be able to sub-
stantiate this claimed performance by field test results.

8.2.6 Condenser Acceptance Testing

Acceptance testing of condensers ought to be done in accordance with the ASME
Power Test code 12.2 [8.103]. Nevertheless, condenser acceptance tests are very
costly and demanding. The test personnel must be experienced and should be super-
vised by a condenser expert. It is furthermore very important to check the appropriate boundary conditions ahead of these tests. For example the air inleakage
rates, the actual performance of the vent pumps, the tightness of the vacuum hoses
connecting the pressure taps to the pressure indication system should be checked,
and the condenser water boxes should be inspected for foreign matter in tubes and
tube sheets.
     Error-analysis furthermore has shown that the ASME PTC 12.2 fouling factor
determination method (§ 8.2.5) is only accurate to within ±30%. On top of that,
the impact of this method on condenser operation and installation costs are considerable. This prompted the development of an alternative measuring technique
that is shown in figure 8.2.25.
     Tubes removed from the condenser immediately before and/or after the conden-
ser tests are cut to lengths of about 1.5 m. Every second tube segment is pickled
Fouled and pickled segments are then installed in two identical test condensers.
The heat transferred in each condenser is controlled towards the same value. The
difference in heat transfer of the two segments is then a true representation of
the effect of fouling. Error analysis has indicated that fouling factors, or
rather fouling resistances, can be determined to within ± 10%.
     While perfecting this fouling test equipment, this error analysis was checked
using materials with different thermal conductivity or different wall thickness
instead of fouled tubes. The difference in heat transfer rates due to differences
in tube wall heat resistance would then have to show up the same way as differences
between fouled and pickled tubes. The heat resistance of a tube can be determined

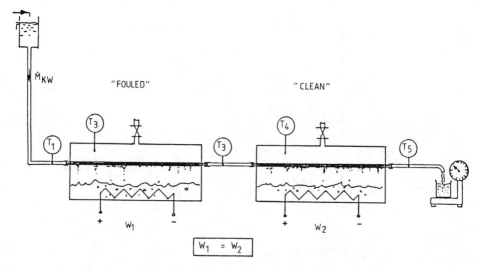

FIGURE 8.2.25   BBC-fouling test rig

from

$$R_t = \frac{d_0}{2\lambda} \cdot \ell n \frac{d_0}{d_i} \tag{8.2.1}$$

where R_t is the resistance, d_0 the outside, d_i the inside tube diameter and λ the heat conductivity of the tube material. Figure 8.2.26 compares the measured resistance with the theoretical resistance according to equation (8.2.1). All but one of the test results lie within the 10% error band and it was confirmed that this one exception was caused by a temporary increase of air leakage into the test condensers.

Temperature stratification was encountered when determining the cooling water outlet temperature according to ASME PTC 12.2 [8.94, § 96]. Experience indicates that at least 3, preferably 5 hydraulic diameters of straight duct should be employed before temperature is measured. Very often this is not practicable, but good results have been obtained instead by using 9 thermocouples of various immersion depth (Fig. 8.2.27).

According to ASME PTC 12.2 [8.94] the condenser pressure is defined as the arithmetic mean pressure in a plane close to the joint between turbine and condenser neck. Because of the possibility of either a pressure gain or a pressure loss in the neck, depending on the shape of the neck and the layout of the extraction piping and the presence of feedheaters, this definition of condenser pressure is not suitable for judging condenser performance.

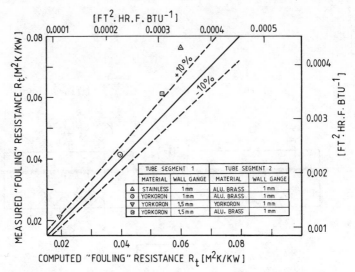

FIGURE 8.2.26 Heat transfer rate differences due to
different tube materials, expressed as heat resistances

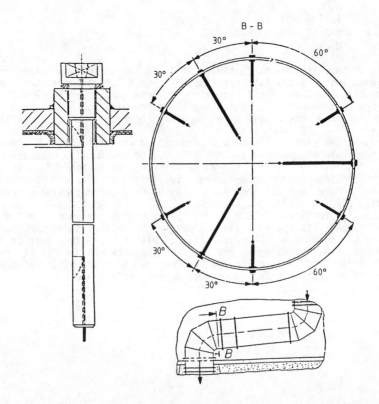

FIGURE 8.2.27 Cooling water outlet temperature measurement

8.2.7   Recent Field Test Results on 'Church-Window' Condensers

Initiated by the author's company as suppliers of these condensers, acceptance tests for some 'church window' condensers were performed parallel to the turbine acceptance tests. ASME PTC 12.2 [8.103] was adhered to with the exception of the method of determining the tube fouling (§ 8.2.5) where an in-house method was employed that was described in § 8.2.6. Data are presented of the four most recent tests in the following power plants :
- Staudinger N° 4, PREAG, 660 MWe
- Setubal N° 1, EdP, 250 MWe
- Ulsan N° 4, KECO, 400 MWe
- Amer N° 8, PNEM, 400 MWe

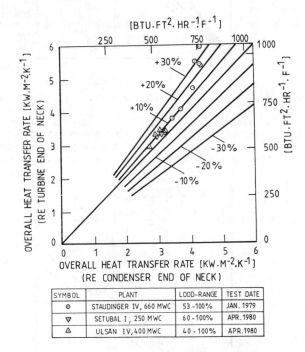

SYMBOL	PLANT	LOOD-RANGE	TEST DATE
⊙	STAUDINGER IV, 660 MWC	53 -100%	JAN. 1979
▽	SETUBAL I , 250 MWC	60 -100%	APR.1980
△	ULSAN IV, 400 MWC	40 - 100%	APR.1980

FIGURE 8.2.28 Effect of pressure measurement plane on heat transfer rates

Figure 8.2.28 shows that the heat transfer rates determined on the basis of the turbine exhaust pressure (defined as condenser pressure by ASME PTC 12.2) always exceed the rates determined on the basis of the condenser pressure at the condenser end of the neck. The magnitude of the difference is dependent on geometries and load. For a representative condenser pressure Brown Boveri therefore recommends taking the pressure at a plane as close as possible to the condenser. The same is recommended by HEI [8.4]. In figure 8.2.29 the measured heat transfer rates based on this representative condenser pressure are compared with predicted rates according to HEI recommendations [8.4]. The latter were obtained using the actual resistance of the fouling. As can be seen, the heat transfer rates of 'church window' condensers are underpredicted by HEI by 5 to 30%. As mention in § 8.2.6, the supplier should best be able to predict the performance of his condenser concept. However, he will be able to do so only by continuous fine tuning of his prediction each time a site measurement is performed. Brown Boveri did so. Unfortunately, these predictions become complicated and involved and have to be performed

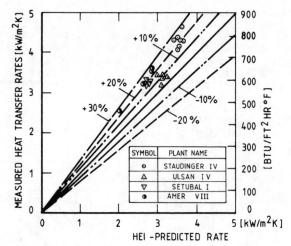

FIGURE 8.2.29 Comparison of field test results with the HEI prediction

on computers. They found that only a prediction method which includes all factors
determining the two phase pressure drop of the steam can be expected to be sufficiently accurate. Figure 8.2.30 shows that the measured heat transfer rates can
then be predicted to within ±5% by this more sophisticated prediction method.

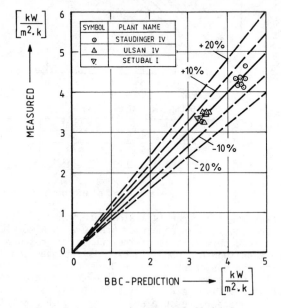

FIGURE 8.2.30 Comparison of field test results
with Brown-Boveri in-house prediction

The measured oxygen concentration in the condensate is shown in figure 8.2.31
for operation with and without makeup water addition. Especially remarkable is the
small increase of the oxygen concentration of only 3 ppb at 40% load in the Ulsan

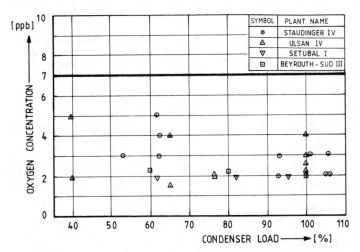

FIGURE 8.2.31 Measured condensate oxygen concentration

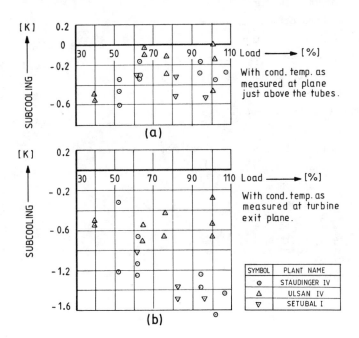

FIGURE 8.2.32 Measured condensate depression

power station during makeup water addition of 17% of the current steam flow rate within the one hour test period. It had never previously been possible to determine the oxygen concentration at loads much below 50% because of the difficulty to maintain stable operation for the required period at such low loads. No deaerating hotwell is employed.

The condensate depression is plotted versus condenser load in figure 8.2.32. Again we are faced with the problem of the definition of condenser pressure. As with heat transfer rates, graph "a" plotted on the basis of the pressure in a plane just above the first condenser tubes has to be considered as representative. As a result 'church window' condensers do not exhibit condensate subcooling but a slight "superheat" (see section 8.2.3.6 for explanation). No reheating hotwell is used in these 'church window' condensers.

NOMENCLATURE

C_p	specific heat
D, D_p	diffusion coefficient
d_i, d_0	tube inside and outside diameter
F	fouling factor
F_x, F_y	distributed forces/unit volume
G	mass velocity
g	acceleration due to gravity
h	depth of tube bundle
k	thermal conductivity
K'	flow resistance factor
K_g	mass transfer coefficient
L	latent heat of vapourization
M	molecular weight
$\dot{m}$	mass flow rate of vapour
p	pressure
Q, q	local condensation rate/unit volume; heat flux
R_0, R_s	universal and steam gas constants
R_t	fouling resistance
r	volume fraction
S	$Pr_c K \left[P_v \mu_v / P_c \mu_c \right]^{1/2}$
S_ψ	sink term, equation 8.1.34
s	tube pitch
T	temperature
U_∞	bulk steam velocity
v_0	suction velocity at the tube wall surface
W	heater input
$\dot{w}$	condensate mass per second
X_{tt}	Lockhart Martinelli parameter
x	vapour quality

Dimensionless Groups

Fr	Froude number, $U_\infty^2 / g d_0$

Ga   Gallileo number, $d_0^3 g / \nu_v^2$

K   phase change number, $L / Cp_c (T_v - T_w)$

Nu   Nusselt number, $\alpha_v d_0 / k_c$

Pe   Peclet number, $U_\infty d_0 / \Gamma$

Pr   Prandtl number, $\mu C_p / k$

Re   Reynolds number, $U_\infty d_0 / \nu_s$

Re_D Reynolds number based on mean free duct velocity

Re_{TP} two phase Reynolds number, $U_\infty d_0 / \nu_c$

Sc   Schmidt number, ν / D

Sh   Sherwood number, $K_g d_0 / \rho D$

Greek symbols

α heat transfer coefficient

β $-v_0 / U_\infty \, Re_v^{1/2}$, equation (8.1.19)

μ dynamic viscosity

ω ratio of free stream air mass fraction to the condensate surface air mass fraction

ζ drag coefficient

ϕ^2 two phase multiplier

ψ dummy dependent variable, equation (8.1.34)

Γ transport coefficient

ρ density

II   Berman PI function, $Fr / (Pr \cdot K)$

σ surface tension

ν kinematic viscosity

Subscripts

acs   air at condensate surface

a   air

c   condensate

cs   condensate surface

cw   cooling water

D   dry, except if stated otherwise

e   exit

gr   gravity controlled

g,G   gas, vapour

G_0 gas only

i   inlet

k   condenser

L   liquid

L_0 liquid only

LMTD   logarithmic temperature difference

N   Nusselt

n   number of tubes

o   overall

s   shellside

sh   shear controlled

scs   steam at condensate surface

v   vapour

w   wall, except if stated otherwise

ws   water side

REFERENCES

8.1   Silver, R.S.: Some Aspects of the Development of the Condenser from the Time of James Watt. *Symposium to Commemorate the Bicentenary of Watt's Patent*, University of Glasgow, UK, September 1969, pp 27-90.

8.2   Sebald, J.F.: A History of Steam Surface Condensers for the Electric Utility Industry. *Heat Transfer Engineering*, Vol. 1, No. 3, Jan-Mar. 1980, pp 80-87.

8.3   Rowe, M.: Power Plant Condensers - Recent CEGB Experience. *Proceedings of Conference on Condensers - Theory and Practice, UMIST*, Manchester, England, 1983.

8.4   Standards for Steam Surface Condensers. *Heat Exchange Institute*, 7th Edition, New York, 1978.

8.5   Recommended Practice for the Design of Surface Type Steam Condensing Plant. *The British Electrical and Allied Manufacturers' Association*, London 1967.

8.6   Orrok, G.A.: The Transmission of Heat in Surface Condensation. *ASME Transact.*, Vol. 32, 1910, pp 1139-1214.

8.7   Guy, H.L. & Winstanley, E.V.: Some Factors in the Design of Surface Condensing Plant. *Proceedings of the Institution of Mechanical Engineers*, 1934.

8.8   Berman, L.D.: Engineering Design of Steam Turbine Condensers. *Teploenergetika*, Vol. 22, No. 10, 1975, pp 34-39.

8.9   Shklover, G.G. & Grigor'ev, V.G.: Calculating the Heat Transfer Coefficient in Steam Turbine Condensers. *Teploenergetika*, Vol. 22, No. 1, 1975, pp 67-71.

8.10 Wenzel, L.A.: in *"Power Condenser Heat Transfer Technology"*, Eds. P.J. Marto & R.H. Nunn. New York, Hemisphere Publishing Corp, 1981, pp 182-183.

8.11 Hopkins, H.L.; Loughead, J.; Monks, C.J.: A Computerised Analysis of Power Condenser Performance Based on Investigation of Condensation. *Proceedings of International Symposium on Condensers, UMIST*, Manchester, England, 1983.

8.12 Nusselt, W.: Die Oberflächen - Kondensation des Wasserdampfes. *VDI Zeitung*, Vol. 60, 1976, pp 541-546 and 569-575.

8.13 Marto, P.J.: Heat Transfer and Two-Phase Flow During Shell-Side Condensation. *Joint ASME-JSME Conference*, Hawaii, March 1983.

8.14 Berman, L.D.: Influence of Vapor Velocity on Heat Transfer with Filmwise Condensation on a Horizontal Tube. *Thermal Engineering*, Vol. 26, 1979, pp 274-278.

8.15 Shekriladze, I.G. & Gomelauri, V.I.: Theoretical Study of Laminar Flow Condensation of Flowing Vapor. *International Journal of Heat and Mass Transfer*, Vol. 9, No. 6, June 1966, pp 581-591.

8.16 Fujii, T.; Uehara, H; Kurata, C.: Laminar Filmwise Condensation of Flowing Vapour on a Horizontal Cylinder. *International Journal of Heat and Mass Transfer*, Vol. 15, No. 2, February 1972, pp 235-246.

8.17 Gaddis, E.S.: Solution of the Two-Phase Boundary Layer Equations for Laminar Flow Condensation of Vapor Flowing Perpendicular to a Horizontal Cylinder. *International Journal of Heat and Mass Transfer*, Vol. 22, No. 3, March 1979, pp 371-382.

8.18 Honda, H. & Fujii, T. in *"Proceedings of the 5th International Heat Transfer Conference"*, Tokyo, Vol. 3, 1974, pp 299-303.

8.19 Lee, W.C. & Rose, J.W.: Film Condensation on a Horizontal Tube - Effect of Vapor Velocity. *Proceedings of the 7th International Heat Transfer Conference*, München, Vol. 5, 1982, pp 101-106.

8.20 Nicol, A.A. & Wallace, D.J.: Condensation with Appreciable Vapor Velocity and Variable Wall Temperature. *NEL Report* 619, 1976, pp 27-38.

8.21 Berman, L.D. & Tumanov, Y.A.: Investigation of the Heat Transfer in the Condensation of Moving Steam on a Horizontal Tube. *Teploenergetika*, Vol. 9, No. 10, 1962, pp 77-83.

8.22 Jakob, M.: *Heat Transfer*, Vol. 1, New York, John Wiley & Sons Inc., 1949, pp 667-673.

8.23 Kern, D.Q.: Mathematical Development of Loading in Horizontal Condensers. *AIChE Journal*, Vol. 4, 1958, pp 157-160.

8.24 Eissenberg, D.M.: An Investigation of the Variables Affecting Steam Condensation on the Outside of a Horizontal Tube Bundle. Ph.D. Thesis, University of Tennessee, December 1972.

8.25 Kutateladze, S.S.; Gogonin, I.I.: Dorokhov, A.R.; Sosunov, V.I.: Film Condensation of Flowing Vapor on a Bundle of Plain Horizontal Tubes. *Thermal Engineering*, Vol. 26, 1979, pp 273-280.

8.26 Fuks, S.N.: Heat Transfer with Condensation of Steam Flowing in a Horizontal Tube Bundle. *Teploenergetika*, Vol. 4, 1957, pp 35-39.

8.27 Wilson, J.L.: The Design of Condensers by Digital Computers. *Institute of Chemical Engineers Symposium Series* No. 35, 1972; also *NEL Report* 619, 1976, pp 132-151.

8.28 Grant, I.D.R. & Osment, B.D.J.: The Effect of Condensate Drainage on Condenser Performance. *NEL Report* 350, April 1968.

8.29 Brickell, G.M.: Potential Problem Areas in Simulating Condenser Performance. In: *"Power Condenser Heat Transfer Technology"*, Eds. P.J. Marto & R.H. Nunn, New York, Hemisphere Publishing Corp., 1981, pp 51-61.

8.30 McNaught, J.M.: Two-Phase Forced Convection Heat Transfer During Condensation on Horizontal Tube Bundles. in *Proceedings of the 7th International Heat Transfer Conference*, München, Vol. 5, 1982, pp 125-131.

8.31 Nobbs, D.W.: The Effect of Downward Vapor Velocity and Inundation on the Condensation Rates on Horizontal Tubes and Tube Banks. Ph.D. Thesis, University of Bristol, UK, April 1975.

8.32 Shklover, G.G. & Buevich, A.V.: The Mechanism of Film Flow with Steam Condensation in Horizontal Bundles of Tubes. *Teploenergetika*, Vol. 25, No. 4, 1978, pp 62-65.

8.33 Fujii, T.: Keynote Address - Condensation in Tube Banks. In: *"Proceedings of the Conference on Condensers - Theory and Practice, UMIST,* Manchester, England, 1983.

8.34 Berman, L.D.: Heat Transfer with Steam Condensation on a Bundle of Horizontal Tubes. *Thermal Engineering*, Vol. 28, 1981, pp 218-224.

8.35 Fujii, T.: Vapour Shear and Condensate Inundation : An Overview. In: *"Power Condenser Heat Transfer Technology"*, Eds. P.J. Marto & R.H. Nunn, New York, Hemisphere Publishing Corp., 1981, pp 193-223.

8.36 Brickell, G.M.: Private Communication, 1979.

8.37 Colburn, A.P. & Hougen, O.A.: Design of Cooler Condensers for Mixtures of Vapors with Noncondensing Gases. *Industrial and Engineering Chemistry*, Vol. 26, 1934, pp 1178-1182.

8.38 Berman, L.D.: Determining the Mass Transfer Coefficient in Calculations on Condensation of Steam Containing Air. *Teploenergetika*, Vol. 16, 1969, pp 68-71.

8.39 Rose, J.W.: Approximate Equations for Forced Convection Condensation in the Presence of a Non-Condensing Gas on a Flat Plate and Horizontal Tube. *International Journal of Heat and Mass Transfer*, Vol. 23, No. 4, April 1980, pp 539-546.

8.40 Koh, J.C.: Laminar Film Condensation of Condensible Gases and Mixtures on a Flat Plate. *Proceedings 4th USA National Congress on Applied Mechanics*, Vol. 2, 1962, pp 1327-1336.

8.41 Lee, W.C. & Rose, J.W.: Comparison of Calculation Methods for Non-Condensing Gas Effects in Condensation on a Horizontal Tube. *Proceedings of International Symposium on Condensers - Theory and Practice, UMIST,* Manchester, England, 1983.

8.42 Berman, L.D. & Fuks, S.N.: Mass Transfer in Condensers with Horizontal Tube When the Steam Contains Air. *Teploenergetika*, Vol. 5, No. 8, 1958, pp 66-74.

8.43 Lee, W.C.: Filmwise Condensation on a Horizontal Tube in the Presence of Forced Convection and Non-Condensing Gas. Ph.D. Thesis, Queen Mary College, London University, 1982.

8.44 Lorenz, J.J. et al.: An Assessment of Heat Transfer Correlations for Turbulent Pipe Flow of Water at Prandtl Numbers of 6.0 and 11.6. *Argonne National Laboratory Memo*, 1982.

8.45 Dittus, F.W. & Boelter, L.M.K.: in *Engineering*, Vol. 2, 1930, p 443.

8.46 Bell, K.J.: in *"Power Condenser Heat Transfer Technology"*, Eds. P.J. Marto & R.H. Nunn, New York, Hemisphere Publishing Corp., 1981, p 86.

8.47 Eagle, A. & Ferguson, R.M.: On the Coefficient of Heat Transfer from the Internal Surface of Tube Walls. *Proceedings of the Royal Society*, Vol. 127, 1930, p 540.

8.48 Sieder, E.N. & Tate, G.E.: Heat Transfer and Pressure Drop of Liquids in Tubes. *Ind. Eng. Chem.*, Vol. 28, 1936, p 1429.

8.49 Sleicher, C.A. & Rouse, M.W.: A Convenient Correlation for Heat Transfer to Constant and Variable Property Fluids in Turbulent Pipe Flow. *International Journal of Heat and Mass Transfer*, Vol. 18, No. 5, May 1975, pp 677-683.

8.50 Zhukauskas, A.A. & Schlancyauskas, A.A.: Heat Emission and Resistance of Checkered Tube Bundles in a Transverse Fluid Flow. *Teploenergetika*, No. 2, 1961, pp 72-75.

8.51 Murray, W.: Private Communication. *CEGB*, NE Region, 1979.

8.52 Butterworth, D.: Modelling of the Gas/Vapor Phase Flow. In: *"Power Condenser Heat Transfer Technology"*, Eds. P.J. Marto & R.H. Nunn, New York, Hemisphere Publishing Corp., 1981, pp 75-81.

8.53 Crossflow Pressure Loss Over Banks of Plain Tubes in Square and Triangular Arrays. Including Effects of Flow Direction. *Engineering Sciences Data Unit* No. 79034, 1979.

8.54 Ishihara, K; Palen, J.W.; Taborek, J.: Critical Review of Correlations for Predicting Two-Phase Flow Pressure Drop Across Tube Banks. *Heat Transfer Engineering*, Vol. 1, No. 3, 1980, pp 23-32.

8.55 Owen, R.G. & Lee, W.C.: A Review of Some Recent Developments in Condensation Theory. *Proceedings of Conference on Condensers - Theory and Practice, UMIST*, Manchester, England, 1983.

8.56 Paten, J.W.; Breber, G.; Taborek, J.: Prediction of Flow Regimes in Tubeside Condensation. *17th National Heat Transfer Conference*, Salt Lake City, 1977.

8.57 Eissenberg, D.M.: Personal Communication to P.J. Marto (Ref. 14), *Oak Ridge National Laboratory*, Oak Ridge, Tennessee, 1977.

8.58 Nicol, A.A.; Aidoun, Z.; Mussa, M.N.: Condensation and Pressure Drop for Crossflow of Steam in Small Tube Bundles. *Proceedings of the 7th International Heat Transfer Conference*, München, Vol. 5, 1982, p 133.

8.59 Lee, N.K.: Simulation of Condenser Pressure Losses by Porous Tubes with Suction. Ph.D. Thesis, University of Bristol, UK, 1981.

8.60 Rowe, M.; Davidson, B.J.; Andrews, E.F.C.; Ferrison, J.A.; Taylor, B.J.: Heat Transfer and Air Blanketing in Steam Condensers. *Conference on Steam Turbines for the 1980s, Institution of Mechanical Engineers*, Paper C180/79, London, 1979.

8.61 Bergelin, O.P.; Bell, J.K.; Leighton, M.D.: Heat Transfer and Fluid Friction During Flow Across Banks of Tubes : VII. Bypassing Between Tube Bundle and Shell. *Chemical Engineering Prog. Symposium*, Series 55, 29, 1959, pp 45-58.

8.62 Bell, K.J.: Final Report of the Co-Operative Program on Shell and Tube Heat Exchangers. *J. Delaware Eng. Exp. Stn. Bull.*, No. 5, Newark, Delaware, 1963.

8.63 Lee, N.K.; Mahew, Y.R.; Hollingsworth, M.A.: Effect of Pitch to Diameter Ratio and By-Pass Lanes on Pressure Loss in Condenser Tube Banks. *Proceedings of International Symposium on Condensers - Theory and Practice, UMIST*, Manchester, England, 1983.

8.64 Chilholm, D. & McFarlane, M.W.: The Prediction of Condenser Performance Using a Digital Computer, *NEL Report* 161, 1964.

8.65 Beckett, G.; Davidson, B.J.; Ferrison, J.A.: The Use of Computer Programs to Improve Condenser Performance. *Proceedings of Conference on Condensers - Theory and Practice, UMIST*, Manchester, England, 1983.

8.66 Barsness, E.J.: Calculation of the Performance of Surface Condensers by Digital Computer. *ASME Paper* 63-PWR-2, 1963.

8.67 Hafford, J.A.: ORCON1 : A Fortran Code for the Calculation of a Steam Condenser of Circular Cross Section. *ORNL TM 4248*, Oak Ridge National Laboratory, Oak Ridge, Tennessee, July 1973.

8.68 Davidson, B.J. & Rowe, M.: Simulation of Power Plant Condenser Performance by Computational Methods. *Naval Postgraduate School*, Condenser Workshop, Montery, California, 1980.

8.69 Harlow, F.H. & Amsden, A.A.: Numerical Calculation of Multi-Phase Fluid Flow. *Journal of Computational Physics*, Vol. 17, No. 1, January 1975, pp 19-52.

8.70 Spalding, D.B.: The Calculation of Free-Convection Phenomenon in Gas-Liquid Mixtures. ICHMT Seminar, Dubrovnik, 1976. Published in *"Turbulent Buoyant Convection"*, Eds. W. Afgan & D.B. Spalding, Washington, Hemisphere Publishing Corp., 1977, pp 569-586.

8.71 Al-Sanea, S.L.; Rhodes, N.; Tatchell, D.G.; Wilkinson, T.S.: A Computer Program for Detailed Calculation of the Flow in Power Station Condensers. *Proceedings of the Conference on Condensers - Theory and Practice, UMIST*, Manchester, England, 1983.

8.72 Al-Sanea, S.L. & Rhodes, N.: Mathematical Modelling of Two Phase Condenser Flows. *2nd International Conference on Multi-Phase Flow*, London, England, 19-21 June 1985.

8.73 Emerson, W.H.: The Application of a Digital Computer to the Design of Surface Condenser. *The Chemical Engineer*, No. 228, No. 5, 1969, pp 178-184.

8.74 Ozeki, T.; Miura, T.; Miyoshi, M.: A Design Method of Condenser Tube Arrangement for Large Power Stations. *Proceedings of Conference on Condensers - Theory and Practice, UMIST*, Manchester, England, 1983.

8.75 Spalding, D.B. (Ed): Developments in Heat and Mass Transfer. *Pergamon Press*, 1976, pp 1183-1188.

8.76 Andrews, J.G.; Deam, R.T.; Smalley, J.: Air Entrainment in Steam Condensers. *Proceedings of Conference on Condensers - Theory and Practice, UMIST*, Manchester, England, 1983.

8.77 Chorin, A.J.: Numerical Solution of the Navier-Stokes Equations. *Math. Comp.*, Vol. 22, 1968, pp 745-762.

8.78 Harlow, F.H. & Welch, J.E.: Numerical calculation of Time-Dependent Viscous Incompressible Flow of Fluid with Free Surface. *Physics of Fluids*, Vol. 8, No. 12, December 1965, pp 2182-2189.

8.79 Patankar, S.V.: Numerical Heat Transfer and Fluid Flow. Hemisphere Publishing Corp., 1980.

8.80 Shida, H.; Kuragasaki, M.; Adachi, T.: On the Numerical Analysis Method of Flow and Heat Transfer in Condensers. *Proceedings of the 7th International Heat Transfer Conference*, München, Vol. 5, 1982, pp 347-352.

8.81 Caremoli, C.: Numerical Computation of Steam Flow in Power Plant Condensers, *Proceedings of Conference on Condensers - Theory and Practice, UMIST*, Manchester, England, 1983.

8.82 Colquhoun-Lee, I. & Bryan, H.G.: Private communication, 1980.

8.83 Webb, R.L.: The Use of Enhanced Surface Geometries in Condensers : An Overview. In: *"Power Condenser Heat Transfer Technology"*, P.J. Marto & R.H. Nunn (Eds.), Washington DC, Hemisphere Publishing Corp., 1981.

8.84 Cooper, J.R. & Rose, J.W.: Condensation Heat-Transfer by Vapor-Side Surface Geometry Modification. *HTFS Report* RS402, 1981.

8.85 Fujii, T.; Honda, H.; Oda, K.: Condensation of Steam on a Horizontal Tube. In: *"Condensation Heat Transfer"*, ASME/AIChE, 1979, pp 35-43.

8.86 Gogonin, I.I. & Dorokhov, A.R.: Heat Transfer from Condensing Freon-21 Vapor Moving Over a Horizontal Tube. *HEAT TRANSFER-Soviet Research*, Vol. 3, No. 6, November-December 1971, pp 157-161.

8.87 Gogonin, I.I. & Dorokhov, A.R.: *J. Appl. Mech. Tech. Phys.*, Vol. 17, No. 2, 1976, pp 252-257.

8.88 Nicol, A.A. & Wallace, D.J.: The Influence of Vapour Shear Force on Condensation on a Cylinder. *Symposium on Multi-Phase Flow Systems*, Institution of Chemical Engineers, Symposium Series No. 38, 1974, Paper D3, pp 1-19.

8.89 Fujii, T. & Oda, K.: *Transactions of Japan Society of Mechanical Engineers* (in Japanese). Vol. 49, 1983.

8.90 Nicol, A.A.; Aidoun, Z.; Mussa, M.N.: Condensation and Pressure Drop for Crossflow of Steam in Small Tube Bundles. *Proceedings of the 7th International Heat Transfer Conference*, München, Vol. 5, 1982, p 133, 1982.

8.91 Anson, D.: Availability of Fossil-Fired Steam Power Plants. *EPRI* FP-422-SR, June 1977.

8.92 Dunham, R.H.: Condenser Tests Prove Value of Code Method. *Power Engineering*, April 1970, pp 42-45.

8.93 Baumann, G. & Oplatka, G.: A New Concept in Condenser Design. *Brown Boveri Review*, Vol. 54, No. 10/11, October-November 1967, pp 327-344.

8.94 Lang, H.: Investigations and Measurements on "Church Window" Condensers, *Brown Boveri Review*, Vol. 60, No. 7/8, July/August 1973, pp 337-344.

8.95 Lang, H. & Oplatka, G.: Theory and Design of "Church-Window" Condensers for Large Steam Turbines. *Brown Boveri Review*, Vol. 60, No. 7/8, July/August 1973, pp 326-336.

8.96 Newson, I.H. & Hodgson, T.K.: The Development of Enhanced Heat Transfer Condenser Tubing. *4th International Symposium on Fresh Water from Sea*, Vol. 1, 1983, pp 69-94.

8.97 Gregorig, R.: Wärmetausch und Wärmetauscher. Aarau und Frankfurt/M., *Verlag Sauerländer*, 1973, pp 274-284.

8.98 High Reliability Condenser Design Study. Stone & Webster Engineering Corp., *EPRI* Research Project RP 1689-10, January 1983.

8.99 Andrieux, B. et al.: Evolution des Condenseurs, des Postes d'Eau et des Séparateurs de Vapeur. *A.I.M.-Liège, Centrales Electriques Modernes*, 1981.

8.100 Bow, W.J. & Wheeler, F.: Surface Condenser Reliability. *Joint Power Generation Conference*, Denver, 1982.

8.101 Sales Documentation, *KWU*, 1.1-6200/5.

8.102 Communication of McHerron to Brown Boveri.

8.103 Steam Condensing Apparatus, *ASME Power Test Codes*, ANSI PTC 12.2, New York, 1975.

Index